ACKNOWLEDGMENTS

This book is the product of more than three decades of cogitation, some of it conscious and directed thinking but much of it a subconscious fermentation of ideas. It would be impossible to determine who, or even how many people, contributed to this process though writings, conversations, or oral presentations. The list of references includes many of them, but one important source not reflected there is Ward Goodenough's 1971 Addison-Wesley module, *Culture, Language, and Society.*

In my struggle to bring some sort of coherence to half-formed ideas, and in an attempt to connect them to work already done by others, I found myself researching fields with which I was either unfamiliar or at best half-familiar. Many of these excursions were directly useful. Others focussed my thinking by proving irrelevant. This rather Darwinian process was fruitful but also slow. I thank Jerry Sabloff for giving me the time to make it work.

A number of people read all or parts of the manuscript, providing me with helpful comments and suggestions. April Nowell, in particular, waded through the entire manuscript, as did Iain Davidson. An anonymous reviewer who did the same provided very valuable suggestions. Jane Kepp provided not only editorial help but suggestions that made the presentation of the material in chapters 1 and 2 much clearer. Janet Monge, Marilyn Norcini, Robert Seyfarth, and Tom Schoenemann all reviewed parts of the manuscript. These people saved me from some embarrassing errors and definitely improved the manuscript. However, especially since I did not always take their advice, they can in no way be held responsible for any of its weaknesses.

Finally, it is largely due to Marilyn Norcini that I had the energy to bring the project to conclusion.

CONTENTS

1

INTRODUCTION

The human way of life is shaped by culture. Culture colors almost everything we perceive, almost everything we think, and almost everything we do. We cannot understand humans without understanding culture, and we cannot understand human evolution without understanding the evolution of culture.

There is a difference – one that seems to have escaped the notice of most investigators – between human culture and anything we may call culture in other species. This is so in spite of many continuities between humans and other primates. The great apes, at least, seem to have most of the cognitive abilities that make human culture possible. Yet there remains a very real and very important difference. Human behavior and ape behavior, like that of all mammals, is guided in part by ideas, concepts, beliefs, etc. that are learned in a social context from other individuals of the same species. Among humans, however, some of these are not just learned socially but are also *created* socially, through the interactions of multiple individuals.

Obviously, I must both explain and defend this statement; I do so briefly in this chapter and in more detail in chapters 2 and 4. The essence of the concept is quite simple. It is, in fact, something that both anthropologists and non-anthropologists probably take more or less for granted in their everyday lives. Yet it has somehow been overlooked by almost all theorists in every discipline dedicated to the evolution of human behavior.

Primatologists often define culture as socially learned behavior or socially transmitted traditions (Alvard 2003; Boesch et al. 1994; Boesch and Tomasello 1998; Laland and Hoppitt 2003; McGrew 1998; Whiten et al. 1999). Archaeological theorists, evolutionary biologists, and sociobiologists have, under rubrics such as memetics and dual inheritance theory, refined this basic concept of culture and applied it to humans (e.g., Boyd and Richerson 1985; Burns and Dietz 1992; Campbell 1965; Cavalli-Sforza and Feldman 1981; Dawkins 1976; 1993; Dennett 1995; Durham 1990; 1991; Giesen 1991; Goodenough 1995; Harms 1996; Rindos 1989; Rose 1998; Wilkins 1998).

Such a model provides a theoretical advantage – or, more accurately, a temptation. If culture consists of particles of behavior or information (often called memes) that are transmitted from one individual to another, then the evolution of culture can be analyzed in terms of natural selection. Cultures evolve when certain memes are more widely adopted than competing memes. Empirically, there is clear evidence that such traditions arise among nonhuman primates (Kawai 1965; McGrew 1998; McGrew et al. 1979; Mertl-Millhollen 2000; Myers Thompson 1994; Nishida 1986; Perry et al. 2003; Van Schaik et al. 2003; Van Schaik and Knott 2001; Whiten et al. 1999; Wrangham et al. 1994). Among humans, there is no question that inventions, ideas, and the like pass from one individual to another. Such a concept of culture therefore makes a good deal of sense.

Among humans, however, there is something quite different that merits the name "culture." This phenomenon is created not by individuals but through interactions among multiple individuals. For example, language (a major part of culture) is the product of many speakers interacting over many generations. Kinship systems are not memes – inventions that each individual is free to accept or reject. As conceptual frameworks, they are created (or maintained or modified) only by multiple individuals through their interactions with one another.

As a result, culture cannot be understood at the level of the individual alone. Knowing the motivations and mental constructs of the individuals involved may be necessary to understand cultural creations or cultural changes, but it is not sufficient. It is also necessary to analyze the interactions of those involved. In this sense, human culture is an *emergent phenomenon* in a way that nonhuman "culture" is not. As Mihata (1997:36) put it,

> what we describe most often as culture is an emergent pattern existing on a separate level of organization and abstraction from the individuals, organizations, beliefs, practices, or cultural objects that constitute it. Culture emerges from the simultaneous interaction of subunits creating meaning (individuals, organizations, etc.)

This emergent property of human culture has important implications. It makes the nature of human social life different in fundamental ways from that of other species (in spite of the continuities that also exist). It makes it possible for groups of humans to coordinate their behavior in ways that are impossible for nonhumans. It changes the relationship of the individual to the social group. Because culture provides motivations for the behavior of the individual, it gives the group a means of controlling the individual that is absent among other primates. Among all living humans, culture provides a (uniquely human) mental or intellectual context for almost everything the individual thinks or does. If culture as an emergent phenomenon is both unique to humans and of major importance to the human way of life, then its origins should be investigated by paleoanthropologists (Paleolithic archaeologists and human paleontologists).

It is my purpose in this book to do four things:

- to analyze and define human culture in a way that will make it possible to investigate its origins
- to propose alternative hypotheses to explain the origins of its various components
- to review the primate evidence to determine to what extent and in what ways culture is unique to humans
- to review the fossil and archaeological data in the hope of identifying the appearance of human culture and in order to test possible alternative hypotheses concerning its origins

I sketch the outline of this process in the remainder of the present chapter. However, the subject is complex, with many ramifications. This chapter offers an idea of where I am going, but it cannot provide a full – or even fully understandable – description of the ideas I am trying to express. This will come only with more detailed discussion in subsequent chapters.

I am under no illusion that I am solving the question of what "culture" is. Some of the best minds in the social sciences and humanities have wrestled with the question and have come to no consensus (Benedict 1934; Boas 1940; Geertz 1973; Kroeber 1952; Kroeber and Kluckhohn 1952; Sapir [in Mandelbaum 1968]; Sahlins 1999; Tylor 1889; White 1949; 1959, to name just a few), and there are even those who argue that the concept should be abandoned altogether (see Borofsky et al. 2001; Fox and King 2002; Trouillot 2002).

What I *am* trying to do is to investigate a particular phenomenon, a particular aspect of the way in which humans govern their behavior, that is different from that of other species. In order to do so, I must have a term by which to refer to the concept I am trying to investigate, and "culture" seems appropriate to me. For other scholars, in other contexts and for other purposes, different concepts will be more meaningful, more useful, or more valid, and the word "culture" will refer to something very different.

To begin with, what I call culture is something that exists in the mind. Several theorists have conceived of culture in this way (e.g., Geertz 1973; Sapir [in Mandelbaum 1968]; Tylor 1889), but my concept of culture is probably closest to that of Ward Goodenough (1981), although it differs from his in other respects. For him, culture consists of categories (forms), propositions, beliefs, values, rules, recipes, customs, and meanings. In a similar vein, when I use the word "culture," I mean something in the mind of the culture bearer that informs and guides his or her behavior.

Behavior and culture are related, but they are not the same thing. Baking a cake is behavior; the recipe followed is culture. A game of football – the interactions among 22 people and a ball – is behavior. The rules that structure that behavior and define it as a game of football are culture.

Of course, culture is not all that exists in the mind and that informs and guides behavior. Such mental coding exists in any animal with a brain, even if the coding is very narrowly determined genetically. Thus hunger, thirst, fear, anger, sexual desire, etc. also help to determine human behavior without being culture.

The same is true of things that are learned by the individual outside a social context. For example, a cat may learn that snow is cold and the armchair by the fire is warm and may shape its behavior accordingly, but these bits of knowledge are not culture. Neither, in my definition, are things that are learned socially but not created socially.

The now famous example of sweet-potato washing by Japanese macaques is a case in point. The practice was invented by one monkey and then learned by other monkeys who observed her (Itani and Nishimura 1973; Kawai 1965; Kawamura 1959; Nishida 1986). Thus the notion of washing sweet potatoes is something that existed in the minds of each of these monkeys. It was learned socially. It guided their behavior. However, it was not *created* through interaction among multiple individuals. It was invented by one monkey, and its creation can therefore be understood in terms of the needs, motivations, and thought processes of a single individual. Even for those monkeys who learned it by observing others, it can be understood in terms of their own individual needs, motivations, and thought processes. It therefore lacks the emergent quality that I attribute to culture.

Thus I use the term "coding" to mean motivations, concepts, beliefs, rules, values, etc., that exist in the mind and that govern behavior. "Culture" is then a subset of coding. The first thing that distinguishes culture from other kinds of coding is that cultural codes are emergent. My concept of emergence is essentially that of complexity theory (e.g., Babloy-antz 1986; Jantsch 1980; Kauffman 1995; Mainzer 1997; Nicolis and Prigogine 1989). That is, emergent phenomena are those that arise from the interactions of multiple agents and that cannot be understood without reference to those interactions.

For example, a football game cannot be understood simply by observing a single football player. It can be understood only in terms of the interactions of all the football players. In this sense, a football game is an emergent phenomenon. However, the game itself is not culture, but behavior. The social (behavioral) interactions of other species are likewise emergent phenomena. In the case of football, however, the behavior of the players is guided by the rules of the game. These rules are themselves emergent phenomena that can be understood only in terms of the interactions of rules committee members, referees, coaches, and players. The coding that produces sweet-potato washing can be understood at the level of the individual alone. The coding that produces a football game cannot. It is therefore culture.

I see absolutely no a priori reason why other species should not have culture in this sense. Yet as will be seen in chapter 4, I can find no good evidence for it in the primatological literature. This is especially striking because the same literature shows that some species seem to have most, if not all, of the necessary cognitive abilities. My statement that culture, as I define it, is unique to humans does not arise from any Cartesian bias. It is an empirical observation and therefore subject to revision in light of new data.

A second important aspect of human culture as it is found among living humans is that its socially created codes provide motivation for behavior. This is not inherent in the nature of socially created coding. Imagine, for example, a population of early humans with simple language (socially created codes for communication) and simple, agreed-upon procedures for cooperative hunts. In this imaginary group, socially created codes would inform and guide the behaviors of the individuals involved, but it would not motivate them. Individuals would hunt cooperatively for the same reasons that other species cooperate: because each individual decided independently that doing so was in his or her own best interest.

However, among modern humans, it appears that culture, in the form of socially created moral beliefs, religious prescriptions, and so forth, motivates behaviors that would be difficult to understand in the absence of culture – for example, celibacy, martyrdom, and wearing a mortarboard and gown while a band plays "Pomp and Circumstance."

If it is in fact the case that culture motivates behavior as well as informs and guides it, then the implications are very significant. It means that the society or social group (however defined) has a way of influencing the behavior of the individual that does not exist in other species. This raises the possibility that an individual might be led to behave in ways that are beneficial to the group yet detrimental to him or her. This in turn raises a theoretical question: how can this happen, given that natural selection should eliminate behavior that decreases the evolutionary fitness of the individual?

This question, usually phrased in terms of the evolution of altruism, is a complex matter that has been the subject of intense investigation. A large body of literature addresses the definition of altruism, the empirical reality of altruism, and theories of group or multilevel selection, as well as a number of related issues (e.g., Aoki 1982; Boorman and Levitt 1980; Brandon and Burian 1984; Chiarelli 1987; Cox et al. 1999; Dugatkin 1999; Field 2001; Frank 1988; Hull 1981; Keller 1998; Maynard Smith 1964; 1976; Pepper and Smuts 2000; Richerson and Boyd 1998; 1999; Smuts 1999; Sober and Wilson 1998; Soltis et al. 1995; D. S. Wilson 1975; Wilson 2002; Wilson and Kniffen 1999; Wynne-Edwards 1962; 1986). How one stands on these issues determines how one is likely to explain the origins of human culture, as I define it. For this reason I discuss the topic in some detail in chapter 3.

The third important characteristic of human culture as we know it today is that it provides a ubiquitous intellectual framework for almost everything we as humans perceive, believe, feel, think, or do. The socially created codes of culture do not replace the older genetically determined or learned codes possessed by other species. We too feel hunger and thirst, we too learn things as individuals outside a social context, and we too learn things by observing the behavior of others, things that we may decide to imitate (or not) depending on our individual motivations.

However, we also live in a world that is full of concepts, definitions, beliefs, values, etc. that are created by culture and that are entirely cultural in their character (Chase 1999; 2001a). We believe in supernatural beings our elders have told us about, we organize ourselves according to social categories that are culturally defined, and we interpret the appearance of a tool, shelter, or item of clothing according to cultural criteria that have nothing to do with its practical effectiveness. We also assign purely cultural meanings to things that exist without culture – to the moon, to sexual desire, and to the bond between mother and child.

Culture replaces nothing, but it incorporates almost everything in a context of culturally defined meanings, values, and beliefs. It becomes a ubiquitous and inescapable framework for everything we perceive, think, or do. Like Geertz (1973:5), I believe "with Max Weber that man is an animal suspended in webs of significance that he himself has spun." These webs are not, however, an a priori consequence of the existence of simple socially created coding. Our imaginary group of humans could very well make use of simple language and practical conventions for co-operative activities without this intellectual superstructure. Thus this ubiquity and all-encompassing character of human culture must also be explained, and its origins traced, if possible, in the archaeological record.

I elaborate on my definition of culture in chapter 2. I also touch in that chapter on some related issues that are not central to the purpose of this book. For example, I discuss briefly the implications of the emergent nature of culture for dual inheritance or memetic analysis, as well as the problem of how an emergent phenomenon such as culture can exist in individual human minds and yet at the same time transcend them to exist at another level. In the remainder of the book, I try to trace and to account for the evolution of culture as a phenomenon.

In doing so, I work from the premise that the three aspects of human culture – socially created codes, motivation by socially created codes, and the elaboration of culture into an all-encompassing phenomenon – may have separate origins. If we assume the contrary, then we will never investigate this possibility, and we risk failing to understand the origins of culture. If, on the other hand, careful investigation indicates that all three are a single phenomenon with a single origin, we will have lost nothing by the effort; in fact, we will have learned something of significance. Clearly, the existence of socially created coding (particularly of language) is a prerequisite for the other two aspects of culture, but it does not necessarily follow that the other two appeared simultaneously with it and in response to the same causes.

In chapter 3, I investigate various possible hypotheses to explain the origins of human culture. It is easy to find adaptive explanations for socially created coding per se. This is especially true since human language is a form of socially created coding. Any adaptive behavior that could benefit from either better communication or better coordination among individuals can serve as a potential explanation for the origins of language. This would include teaching one's offspring verbally, rather than having them learn only by observation and imitation. It would include cooperative activities such as hunting. It would also include behaviors not found in other mammals. For example, a group might enhance its chances of finding food by dividing into several small foraging parties, agreeing to meet at a specific location and share either food or information.

I propose a series of alternative hypotheses to explain how culture came to provide motivation *and* how culture came to be an all-encompassing system. These include

1. The hypothesis that culture is a by-product of simple socially created coding

2. Hypotheses that explain culture in terms of psychological bene-
    fits for the individual
3. Hypotheses that explain culture in terms of group benefits

I also discuss the archaeological test implications of these hypotheses.

In chapter 4, I investigate the origins of simple socially created coding. The first task is to review the primatological record for field or laboratory evidence that primates other than humans construct, through social
interaction, codes that govern the behavior of individuals. In fact, the
data seem to be remarkable in this respect. Nonhuman primates, particularly the great apes (chimpanzees, bonobos, gorillas, and orangutans)
appear to have almost all the cognitive abilities needed to construct and
make use of such codes.

Yet there is no good evidence that they actually do so. It is not a part
of their adaptation in the wild, and even in the laboratory, for all their
apparent symbolic abilities, they seem to stop just short of doing so. This
conclusion is subject to change, of course, as further evidence is collected. But for the time being it appears that, whatever the cognitive
abilities of our nearest relatives, all three aspects of human culture
evolved after our lineage separated from theirs.

The fossil record provides two kinds of evidence for the origins of
language. Language, of course, is a set of socially created conventions
for communication, and as such it is a major part of culture. Human paleontologists and neuroanatomists have used endocasts of fossil crania
and reconstructions of vocal tract anatomy to try to trace the origins of
language. I argue in chapter 4 that there are crucial gaps in the chains of
argument, inference, and data that link either set of evidence to the origins of language, and that at present neither provides conclusive evidence about the origins of socially created coding. Nevertheless, the evidence from both is suggestive of an origin for language before the end of
the Middle Pleistocene.

In the same chapter, I argue that most archaeological evidence for
symbolism tells us little about the earliest origins of socially constructed
coding. On the one hand, there are significant weaknesses in arguments
linking the origins of stone tool making per se to the origins of language.
On the other hand, the best archaeological evidence for socially created
coding is actually indirect. If socially created coding predates the expansion of culture into systems of religion, mythology, and ritual, then archaeological evidence for these will postdate the origins of socially created coding.

The reason is that a population (such as the imaginary one introduced
earlier) that makes use of socially created coding in a limited way for
specific practical purposes would have no reason to produce symbolic
artifacts or to use artifacts to express cultural meaning. Thus their mate-

rial "culture" will look much like that of intelligent populations without socially created coding. Archaeologically, the best evidence for the origins of this phenomenon will be direct evidence of complex cooperative behavior. There is evidence from the site of La Cotte-de-Saint-Brelade for mammoth drives by the late Middle Pleistocene (Scott 1980), and evidence from a number of European sites for drives of large ungulates in the early Upper Pleistocene (David and Fosse 1999; Farizy et al. 1994; Hoffecker et al. 1991; Jaubert et al. 1990; Klein 1979; 1987; Klein and Cruz-Uribe 1996; Levine 1983). These data are in line with suggestions from the fossil record that the origins of socially created coding lie at least as far back as the Middle Pleistocene.

In chapter 5, I apply what we know about the archaeological record to test the alternative hypotheses proposed to explain why cultural coding motivates individual behavior and why culture became an all-encompassing phenomenon. The most relevant aspects of the archaeological record are evidence for ritual, mythology, and religion (in the forms of art, use of coloring, and musical instruments). Of necessity I address the serious taphonomic and epistemological problems involved in interpreting the archaeological record (Barham 2002; Bar-Yosef 1988; Bednarik 1992; 1995; Belfer-Cohen and Hovers 1992; Chase 1991; Chase and Dibble 1987; 1992; Deacon 2001; D'Errico and Villa 1997; Duff et al. 1992; Gargett 1989; Gowlett 1996; Klein 2000; Marks et al. 2001; Marshack 1976; 1989; 1990; McBrearty and Brooks 2000; Mellars 1991; 1996a; Mithen 1996b; Noble and Davidson 1996; White 1992; Wurz 1999; Wynn 1996). Style and artifact standardization, in particular, are concepts whose relationship to the question of culture as I define it have not been clearly worked out in the Paleolithic archaeological literature.

In the end, it seems to me, judging from currently available data, that the ubiquitous and all-encompassing nature of human culture is probably a mechanism by which socially created coding can be used to motivate and influence the behavior of individuals for the benefit of the larger social group. This implies that genetic evolution has not produced fully altruistic humans. However, it also implies that genetic evolution has in one way or another produced humans who are, to an extent, willing to let socially created codes, codes that are external to us as individuals, motivate our behavior. Culture and genetics work together to produce the human way of life.

All these conclusions are to some extent tentative. This is in part because no one has ever explicitly set out to investigate the evolution of human culture as I conceive of it. As a result, the empirical research on which conclusions must rest was designed with other ends in mind. Yet it is also true that science is continuously working at the edges of what is

known, and that scientists must base their work on imperfectly understood or imperfectly known foundations. I assess the state of our knowledge in the concluding chapter.

It should be understood from the beginning that I have no expectation that my conclusions will stand forever. I expect that even my list of
alternative hypotheses, my analysis of test implications, and my analyses
of how to interpret the data will be challenged. My purpose in this book
is not to provide final answers. Rather, it is to raise the issue of the evolution of human culture as an emergent, socially created phenomenon
and to make it a part of the research agenda for Paleolithic archaeology.

2

HOW IS HUMAN CULTURE DIFFERENT?

In the first chapter, I argued that human culture is different from any-thing found in other species, and I outlined, briefly, my idea of what hu-man culture is. In this chapter, I explain the concept of human culture in more detail. Before doing so, it will be useful to offer a reminder of just how different the human way of life is from that of other species. The difference is qualitative, not just a matter of degree.

This position is not a theoretical one but a matter of empirical obser-vation. Darwin's *The Origin of Species* situated humans squarely within nature, established that we are animals, and demonstrated that our spe-cies is related to all others both by nature and by descent. These findings have been amply confirmed by a huge body of scientific research carried out since the book's publication. Yet at the same time it is clear that hu-man behavior differs in important respects from that of other animals.

This is not a contradiction. The observation that humans are in some ways distinct implies no rejection of our material nature, no a priori Car-tesian philosophical bias. Every species must be unique in some respects, or separate species would not exist. Every species shares some of its traits with all animals, and other traits only with closely related species, but in the end some trait or traits will distinguish each species from even its nearest relatives.

Thus the fact of human uniqueness is not in itself remarkable. Yet our species has chosen a rather peculiar way to be unique. In the course of our evolution, we have done more than change our anatomy, physiol-ogy, and behavior. We have also changed, in part, the manner in which our behavior is governed.

Humans are primates, and for the most part we do essentially what other primates do. In many cases where we differ, the difference is one of degree rather than of kind. For example, it has been suggested that at least some apes have all the abilities needed to use symbolic language, albeit in less developed form than humans (see Savage-Rumbaugh et al.

1998:77-138 for an especially vigorous statement). In spite of this, it is easy to find things done by humans that other living species simply never do. Let me give three examples.

In the nineteenth century an unusual group flourished in the United States, the United Society of Believers in Christ's Second Appearing, better known as the Shakers. From an evolutionary perspective, the most remarkable thing about this group is that it was adamantly celibate. This celibacy included all Shakers, not just religious specialists such as priests or nuns. Since evolutionary success is synonymous with reproductive success, such behavior is difficult to explain. In fact, it is so rare that it seems to be confined to humans. It is certainly difficult to imagine a chimpanzee accepting a life of celibacy.

What is most human about the behavior of the Shakers, however, is not the fact of celibacy but the reasons for it. Shakers perceived all sexual relations as spiritual pollution or worse: "Every marriage, however proper for the *world and its children,* crucifies Christ afresh; every sexual congress of the twain, however necessary for the peopling of the earth, pollutes the *Christian temple*" (*Manifesto* 8 [1878]:43, in Collins [2001, emphasis in the original]).

This attitude was rooted in the Shakers' concept of the spirit and the flesh and in their reading of the Bible. In the *Testimony of Christ's Second Appearing,* published by order of the Ministry of the Society, the serpent of the Garden of Eden is equated with the devil, and lust, with the serpent's head, which was the serpent's superior part, "...his highest affection; that in which he finds the most supreme delight" (Youngs 1810:46-48).

> And such is that feeling and affection, which is formed by the near relation and tie between the male and female; and which being corrupted by the subversion of the original law of God, converted that which in the beginning was pure and lovely, into the poison of the serpent; and the noblest affection of man, into the seat of human corruption. (Youngs 1810:48-49)

Shakers behaved as they did because their actions were governed by a religious worldview, a set of concepts, values, and beliefs the like of which would be utterly foreign to any other species. It is inconceivable that members of any other species would remain celibate because of theological philosophy.

The game of chess is another example of how different humans are, in certain respects, from other primates. Competition and play are virtually universal among mammals. Games like chess are not. Not only is chess based on arbitrary conventions having nothing to do with the "real" world, but the concept of chess is itself pure convention. Other species compete, and other species know when one party to a competition has won or lost, but winning or losing is a down-to-earth matter involving physical force, territory, access to mates, and the like. No other species would define winning and losing as arbitrarily and abstractly as the International Chess Federation does in its rules:

> Article 9: Check
> 9.1. The king is in "check" when the square it occupies is attacked by one or more of the opponent's pieces; in this case, the latter is/are said to be "checking" the king. A player may not make a move which leaves his king on a square attacked by any of his opponent's pieces.
> 9.2. Check must be parried by the move immediately following. If any check cannot be parried, the king is said to be "checkmated" ("mated").
> 9.3. Declaring a check is not obligatory.

And the rules go on and on – the definitions of defeat, stalemate, and draw continue for a further 16 subarticles of Article 10.

Finally, I cannot resist including a remarkable example of something that must be considered uniquely human, the fact that we create, discuss, and take seriously fictional worlds that we know very well do not really exist. The following excerpt is from a World Wide Web site dedicated to the language of the fictional Klingons in the *Star Trek* television series:

> In operation since 1992, the Klingon Language Institute continues its mission of bringing together individuals interested in the study of Klingon linguistics and culture, and providing a forum for discussion and the exchange of ideas. ... The Klingon Language Institute is a nonprofit 501(c)3 corporation and exists to facilitate the scholarly exploration of the Klingon language and culture. (From the Web site of the Klingon Language Institute, http://www.kli.org/kli/, June 2, 2003)

The following exchange took place on a Web site dedicated to *Star Trek* discussions. It concerns the facial morphology of Klingon characters:

> *Tribble565 (6/7/02):* what is it with the cranial ridges and how in the heck did Kang Koloth and Kor change to have them [I'm] still waiting for a reply
> *Frogden (8/14/02):* The new look Klingon derived from the need to dramatize the facial features, to appear more evil. There was some official explanation which I don't recall, but as it is only fiction, does it really require explanation?
> *Tribble565 (12/22/02):* Yes it matters you freaking idiot. To sci-fi fans just because something isn't real doesn't mean that they don't require a real explanation. (From SJ's Realm Forums,

++http://pub40.ezboard.com/fsjsrealmforums-
startrek.showMessage?topicID=41.topic, June 2, 2003)

It is difficult to imagine a chimpanzee becoming equally concerned about the anatomy of purely fictional beings.

The reason I cite these examples is to emphasize what everyone already knows but sometimes tends to forget: that humans think and behave in ways that other animals do not. Chimpanzees do not practice celibacy for doctrinal reasons, do not play games like chess, and do not invent and discuss fictional worlds. These differences are not differences of degree. In spite of all the other continuities between us, in ways such as these other primates simply do not behave or think as we do. Why this is so is the crux of the issue. I will argue that it is not because we are more intelligent, although intelligence is important, but because our way of life is shaped by culture.

Recall that I use the term "culture" to refer to the totality of three related phenomena:

1. Codes that we create through social interaction inform and govern our behavior. These codes are emergent in character because they cannot be understood without reference to this interaction. The codes do not replace other, private, forms of coding, but are added to them.
2. Such socially created codes not only inform and govern our behavior but also frequently *motivate* it. Because this potentially leaves individuals open to exploitation by the social group that creates the coding, our willingness to be motivated by socially created coding can be seen as a susceptibility to cultural manipulation.
3. Cultural codes form all-encompassing webs of meanings, values, and dicta that incorporate into themselves almost everything that humans perceive, think, or do. Thus culture forms an inescapable intellectual framework for human life and human action.

The heart of this chapter is a detailed explanation of what I mean by each of these phenomena. Once this has been accomplished, I flesh out my concept of culture by explaining how I see it operating in the normal course of human life. Finally, I touch briefly on two implications of my characterization of culture that, while not directly related to the subject matter of this book, are nevertheless of some interest: its implications in terms of complexity theory, and its implications for the concept of culture as a superorganic phenomenon.

2.1. SOCIALLY CREATED CODING

My concept of coding is an expansion of the dichotomy between genotype and phenotype. Thus coding stands in the same relationship to behavior that the genotype stands in relationship to the phenotype. By coding, however, I mean something that exists in the mind (or brain) that governs and informs behavior.

We can think of coding in terms of four categories or levels:

1. Coding that is essentially determined genetically. Note that, like all coding, this is something in the brain, not the behavior it produces.
2. Learned coding. Because of the plasticity of their brains, mammals are able to create new codes in response to their interactions with their environments.
3. Socially learned codes. These codes are initially created by one individual through individual learning, but others then learn them from conspecifics, either by observation or through teaching.
4. Codes created through social interaction.

In vertebrates, the coding that governs behavior is located in the brain. The brain works by the movement of electrical impulses through networks of neurons. The topology of these networks and the chemical states of the synapses, or connections between them, determine how sensory input is translated into motor output or behavior. We need not go into any detail concerning this process. It is sufficient to say that there are neural structures in the brain that determine how an animal will behave in the presence of given sets of external and internal stimuli. Essentially, it is these structures that I call "coding."

The concept of coding should not be understood narrowly, as referring only to stimulus-response operations or just to rules or algorithms for behavior. Consider what must happen if a cat is to catch and eat a mouse. It must feel hungry. It must have some idea of what a mouse is and that eating a mouse will satisfy its hunger. It must go to where a mouse is likely to be found. It must search for mice, and when it sees, hears, or smells one, it must "recognize" the sound, sight, or odor as indicative of something edible. It must stalk the mouse, spring on it, seize it, kill it, and eat it. In the whole process, it must be able to walk across uneven terrain, keeping its balance and moving its limbs appropriately. It must be able to coordinate its vision, sense of balance, and sense of where its own body parts are so that when it springs it will land on the mouse. If it sees a dog approach, it must abandon its hunt and climb a tree. I include in the concept of coding everything in the brain of the cat that makes these things possible.* This would include sensations, emotions, motivations, knowledge, memories, categories or concepts, rules or algorithms, and much more. Thus, what I mean by coding is very broad in scope.

The relationship between coding and behavior is not rigidly fixed. An animal's behavior will depend on how the coding of the brain processes all the external and internal stimuli in a given situation, so that the end result is the product as much of circumstances as of the neural coding itself. In addition, different codes may compete for control of an animal's behavior. (This will be important to remember when we come to cultural codes.)

Consider a cat that is both tired and hungry. The sensations of hunger and of fatigue are neural codes that motivate it to behave in certain ways, but it is by no means certain how this cat will act. It may remain where it is, resting; it may go hunting; it may go hunting, but in a half-hearted, lackadaisical manner; and so forth. The same can be said about a cat that is both hungry and afraid or about a cat that is tired, hungry, and afraid. In other words, to say that there is coding that motivates a cat with an empty stomach to hunt is not to say that a cat with an empty stomach will necessarily go hunting. Rather, the cat's behavior will depend on interaction and competition among multiple codings in the context of a specific set of external circumstances and internal conditions.

2.1.1. Noncultural Coding

2.1.1.1. Learning

Although genetics plays a key role in the construction of the brain, both environment and experience shape the brain during a young animal's development. In other words, the actual forms or characteristics of the neural structures of the brain are determined in part by environmental factors and by the animal's experiences during growth and development.

In all mammals, the brain continues to change in response to external stimuli during the entire lifetime of the individual. We are, in fact, genetically coded to be able to rework our neural coding. This plasticity of the brain and its neural structures – "learning," in ordinary language – is a major part of the adaptive strategy of mammals.

Different kinds of neural structures exhibit different degrees of plasticity. Some functions are almost completely fixed, at least by adulthood. Examples include the sensation of pain in response to injury, color per-

---

* I deliberately sidestep the philosophical controversy about the relationship between "mind" and "brain" on the grounds that it is essentially irrelevant to my purposes in this book.

ception, and "knowing" where our limbs are even when we cannot see them. Other neural codes are partially plastic. Breathing is something we know how to do at birth, but a diver or musician can learn new ways of breathing. Still other coding is extremely plastic. For example, mammals are constantly learning spatial information. A pet cat learns when and where it gets fed, where its litter box is, and where the cat door is located. If its owner moves the litter box, the cat will learn the new location. Even an adult cat can learn a new algorithm or skill, such as how to use a cat door.

Thus learning is the modification of neural structures in order to create new codes or to modify existing ones. This involves an interaction between the environment and existing codes. New codes will be created that, in general, fit with existing ones. In other words, an animal will learn to do something that satisfies existing codes (e.g., hunger) and to avoid behaviors that do the opposite (e.g., eating foods that cause nausea). Both genetically determined and learned neural coding are involved. If an interaction with humans causes an animal pain (genetically based coding), that animal will learn to fear humans (both genetic and learned coding) and will therefore be reluctant to eat food that is too near a human, even when the animal is hungry. Extreme hunger may outweigh this fear, so that the animal may feed near humans. If no one bothers it and it can satisfy its hunger often enough, it will eventually unlearn its fear of humans.

The borderline between learned and genetically determined coding is not only blurred but also complex. First, nothing can be learned unless the requisite neural structures are present. This means that the kinds of things that can be learned by members of a given species is genetically delimited. A reptile cannot learn human language, for example. At the same time, there may be specialized, genetically coded neural structures for learning specific kinds of information or skills. For example, humans seem to have specialized neural structures for recognizing human faces (Alcock 2001:171-174) and perhaps for categorizing living things (Atran 1990; Herrnstein et al. 1985; Poole and Lander 1971).

In addition, there are many skills that seem to be genetically determined because under normal circumstances all members of a species learn them, yet they must be learned. Humans, for example, must learn to walk bipedally, and songbirds must learn the songs appropriate to their species (Marler and Tamura 1964).

The relationship between genetics and learning is both interesting and, in a general sense, important – but it is of little relevance to the present discussion. My main point here is to explain what I mean by neural coding. The crux of my argument depends not on the difference between genetically determined and learned coding, but on the difference between individual coding and socially constructed coding.

Note that both inbred and learned coding (and everything in the gray area between them) is particular to the individual animal. Granted, codes may be "shared" in the same sense that blue eyes may be "shared" by two individuals. More than one individual may have similar neural structures for perceiving colors, and more than one individual may have learned that a certain food tastes good. However, these individuals do not actually share the same eyes or neural code. Each has a copy, but each copy is physically distinct and internal to the individual organism. Most important of all, the creation of each copy is in a sense particular to the individual. Learned codes are created by each individual interacting with its environment. Even if the neural structures or the behaviors they produce are similar, each individual animal must nevertheless create the codes for itself.

I emphasize this private nature of learned codes because, as I will explain, cultural codes differ fundamentally in that they are created, maintained, and modified publicly by the interactions of multiple individuals.

2.1.1.2. Socially Learned Coding

Animals, then, learn by interacting with their environment. Other individuals of the same species constitute an integral part of an animal's environment, and members of at least some species are capable of learning by observing the behavior of conspecifics. As a result, something learned independently by one individual may spread through a population when others observe the first individual. To many scholars, this is the essence and the definition of culture (e.g., Alvard 2003; Boesch et al. 1994; Boesch and Tomasello 1998; Laland and Hoppitt 2003; McGrew 1998; Whiten et al. 1999). In my opinion, something more is going on among humans. Learning from conspecifics is an important part of human culture, but it is not the whole picture.

There are famous examples of socially learned coding among nonhuman species. In three species of tits (*Parus*), individual birds learned from others about opening milk bottles (Fisher and Hinde 1949). They either removed or broke through the cardboard caps of milk bottles to drink the cream and milk inside. Several lines of evidence indicate that this trick was not discovered individually by each tit, but that there were "pioneers" and learners. Apparently, more than one bird independently discovered this manner of obtaining nourishment. Often, an increasingly large portion of the local tit population would then learn and adopt the practice of opening milk bottles.

Another famous example is the washing of sweet potatoes by a troop of monkeys (*Macaca fuscata*) on the Japanese islet of Koshima (Itani and Nishimura 1973; Kawai 1965; Kawamura 1959; Nishida 1986). The troop was a wild population but was being provisioned with food. In 1953, one young female named Imo began washing sweet potatoes in a stream, presumably to remove sand. The practice was learned by a close peer of hers and then by other young monkeys and by older monkeys closely related to Imo. Offspring of females who washed sweet potatoes learned the habit from their mothers, and the practice became widespread among all but the oldest members of the troop.

In 1956, Imo discovered a way of separating wheat from the beach sand where the human providers placed it. She would throw a handful of wheat and sand into the water. The wheat would float, and she could scoop it up. This innovation, too, spread to many members of the troop. (For a more skeptical view of this example, see Tomasello 1999:519).

Such learned traditions are common among chimpanzees. Nine chimpanzee ethologists recently compiled a database of behaviors observed in different parts of Africa (Whiten et al. 1999; see also Nishida et al. 2004). They listed 39 behaviors that were customary or habitual in some areas but absent from others, behaviors for which they could find no environmental explanation. These included fishing for ants, using a hammer and anvil to crack nuts, tickling oneself with an object, and clasping one's arms overhead during grooming. These behaviors had not been invented independently by each individual chimpanzee, because in that case they would not have been common in some areas and absent in others.

Similar patterns of variation have been observed among bonobos (Hohman and Fruth 2003). Orangutans in some areas use tools to feed on *Neesia* fruits. In others, they do not, and these differences also seem indicate learned traditions (Van Schaik et al. 2003).

In fact, learning in a social context is almost inevitable among animals for whom learning is an important part of their adaptation and who are also dependent on adults during their infancy. This produces traditions that are perhaps less spectacular than milk-bottle raiding or sweet-potato washing but that are learned traditions nevertheless. Avital and Jablonka (2000:105-107) vividly described one such tradition or set of traditions:

> Dusk is a good feeding time for village mice. The small, four-month-old, grayish brown female domestic mouse silently scales the outer wall of the village grocer's warehouse. She enters the warehouse through a small crack in the wall, and quickly slides down to the piles of bags containing pinhead oatmeal and canary seed. This urine-marked route leads safely to the best source of solid food around. It

was first introduced to her by her mother, three months ago, and has been used by her ever since, at least twice a day, at dawn and dusk.... Her scent survey discovers no rats, cats or strange mice, so she can now safely dive into one of the bags of oatmeal and eat as much as two grams, almost a quarter of her own weight. The pinhead oatmeal is always her first choice. But why? Mice are omnivorous and will eat almost anything, and canary seed is a well-known mouse delicacy; but, like every other mouse, this doe has some loyalty to the first solid food she ever smelled and tasted. In her case it was the oatmeal of this warehouse. (p. 105)

After her young are born and old enough to introduce to the outside world, she

leads a group of stiff-haired, hesitant youngsters up the red brick wall on their way to the warehouse. Suddenly a strong smell reaches their sensitive muzzles, the smell of a brown rat, a notorious mouse-hunter. In a split second the alarmed mother changes direction and leads a scampering group back to tool shed, nest and safety. The youngsters will remember the traumatic smell of the rat for a long time, and know what to do when they smell it again. At dusk, the same team tries again and succeeds, this time without trouble, in entering the warehouse via the well-trodden urine-marked route, and enjoys the pinhead oatmeal. From now on, the warehouse feeding site, and the special routes leading to and from it, will be the youngsters' first choices. (p. 107)

In other words, in a species in which learning leads to individual differences in knowledge and behavior, family traditions arise to the extent that young animals learn from observing or even just accompanying their mothers.

Individuals learn from interactions with their environment. In social learning, they learn by observing one part of that environment, the behavior of conspecifics. One individual creates a new code (i.e., learns something), such as opening milk bottles to get at the milk or cream inside. Other individuals observe this first individual's behavior and then use their observations to create codes in their own brains that produce the same or similar behavior.

Such social learning produces a phenomenon analogous to genetic evolution. The recognition of this fact makes possible a theoretical stance in which (1) human and much animal behavior is considered to be the product of dual inheritance (genetic and cultural), and (2) cultural evolution is seen as an essentially Darwinian process.

In this view, when behaviors are learned from conspecifics, they are replicated in a manner that is essentially equivalent to the replication of genes. The more individuals who learn a new behavior or a new bit of knowledge, the more copies of that behavior or item of knowledge exist in the population. Dawkins (1976:192) recognized this similarity and coined the word "meme" to refer to such replicated units of culture. He chose the term deliberately to emphasize the analogy to the gene, which is the replicator in biological evolution.

This parallel between the transmission and selection of genes and the transmission and selection of memes has inspired a vast and influential literature that analyzes both human and nonhuman behavior and the codes that produce that behavior in terms of memes (e.g., Boyd and Richerson 1985; Burns and Dietz 1992; Campbell 1965; Cavalli-Sforza and Feldman 1981; Dawkins 1976; 1993; Dennett 1995; Durham 1990; 1991; Giesen 1991; Goodenough 1995; Harms 1996; Rindos 1985; 1989; Rose 1998; Shennon 2003; Wilkins 1998). This literature encompasses parts of evolutionary biology, sociobiology, and archaeological theory.

It is not my purpose to review or to critique this work, which goes under rubrics such as memetic evolution, dual inheritance, and cultural selectionism. However, I must briefly expound its basic outlines in order to show where human culture departs from the memetic model. (For the sake of convenience, I use terms such as "meme" and "memetic" as a shorthand for socially learned codes and their transmission.)

Darwinian evolutionary theory, of course, demands selection as well as replication, and selection does in fact operate in memetic traditions. Not all memes will be replicated as frequently as others. In genetic evolution, an allele is replicated when it is passed on to a viable son or daughter. The more often this happens, the more copies of an allele exist in the gene pool of a population. Thus (genetic) evolutionary selection depends on how many viable offspring the bearers of a given allele leave behind, and differential reproduction lies at the heart of competition between alleles.

The process is very similar in memetic evolution. Replication consists of the learning or adoption of a meme by a new individual. The more individuals who adopt a meme, the more copies of that meme there will be in the "meme pool" of a population. Competition between memes is based on how many individuals adopt each meme.

Of course, the process is not entirely analogous to biological evolution. What causes an individual to accept or reject a meme is, essentially, how well or how poorly it fits with that individual's already existing coding. The same is true of nonsocial learning:

> In a variable environment, it is clearly useful to be able to develop the locally adaptive phenotype. But how does the organism determine what that phenotype might be? There are many ways, but in most species these processes share the same general features. The organism inherits criteria that determine what feels good and what feels bad; feelings of security and satiation are good, and feelings of fear

> and hunger are bad. ... The organism tries a variety of behaviors and
> retains those which are associated with rewarding sensations. In this
> way, complex patterns of behavior appropriate to local conditions can
> be generated. (Boyd and Richerson 1985:14)

A tit, then, will attempt to open a milk bottle if the behavior of another tit leads it to believe that doing so will satisfy its hunger. It will not
do so if, because of some previous experience, it does not believe this, or
if a fear of humans keeps it from approaching houses. Thus the preexisting neural codes that determine whether or not an animal will accept a
given meme consist of those that are genetically determined (e.g., hunger), those that have been learned independently (e.g., the taste of a certain kind of food), and perhaps even those that have been learned from
another individual or individuals (e.g., fear of humans).

Essentially, memetic or socially learned coding resembles individually learned coding in that each individual animal creates its own codes.
In the individual case, it creates codes in response to its own direct interactions with its environment. In the memetic case, it does so after observing the behavior of other individuals – and this means that memes are
replicated. How often a meme is replicated depends on how many individuals have preexisting neural coding that leads them to adopt that
meme. Therefore the successful meme is one that adapts not to the
physical environment but to the existing pool of neural coding in a population. The locus of memetic selection is the neural coding of the individual.

There are other differences between genetic and memetic evolution
that are of less interest to us here (Rose 1998; Tracy 1996; Weiss and
Hayashida 2002; see also Daly 1982). Genes are indubitably coding, not
behavior. In the case of memes, this is much less clear. If a mother explains to a child how to do something, then a code is being replicated;
but if the child learns a behavior through observation, then the behavior,
not a code, is replicated. In either case, however, the individual adopts a
meme by creating its own internal neural coding, just as it does when it
learns something on its own. In no case is a code transmitted physically
from one individual to another, as in genetic replication.

Among humans, of course, memes may be transmitted by deliberate
teaching and by means of language. Deliberate teaching has also been
claimed for some nonhuman primates as well, albeit on a much smaller
scale (Boesch 1993; King 1999). I will come back to this claim in chapter 4.

Among humans, there are certainly codes that resemble memes.
Dennet (1995:344) gave a list of examples:

These new replicators are, roughly, ideas. Not the "simple ideas" of Locke and Hume (the idea of red, or the idea of round or hot or cold), but the sort of complex ideas that form themselves into *distinct memorable units* – such as the ideas of

arch
wheel
wearing clothes
vendetta
right triangle
alphabet calendar
the Odyssey
calculus
chess
perspective drawing
evolution by natural selection
impressionism
"Greensleeves"

Dawkins (1976:192-193) gave another example in the chapter in which he coined the term "meme":

Consider the idea of God. We do not know how it arose in the meme pool. Probably it originated many times by independent 'mutation.' In any case, it is very old indeed. How does it replicate itself? By the spoken and written word, aided by great music and by great art. Why does it have such high survival value? Remember that 'survival value' here does not mean value for a gene in a gene pool, but value for a meme in a meme pool. The question really means: What is it about the idea of a god that gives it its stability and penetrance in the cultural environment? The survival value of the god meme in the meme pool results from its great psychological appeal. It provides a superficially plausible answer to deep and troubling questions about existence.

There is certainly something very meme-like about all these ideas. They are indeed learned socially from other humans. Some such ideas survive and spread; others die out. Thus meme-like entities are common to both humans and other species. For most of the scholars whom I have cited, "culture," including human culture, is synonymous with the social learning of particles of either behavior or coding.

However, I believe there is an element to human culture – and to most if not all of the examples just listed – that goes beyond and sets it apart from the memes found in other species. Social codes are not just transmitted from one individual to another; they are *created* by interactions among individuals. This makes cultural codes, unlike memes, *emergent* phenomena. Many of the meme-like ideas listed above differ in this respect from memes found in other species.

2.1.2. Emergence

In order to explain the foregoing, I must clarify what I mean by emergence. There is nothing complicated about the concept, but it is essential for understanding what I have to say.

> The concept of emergence is most often used today to refer to the process by which patterns or global-level structures arise from interactive local-level processes. This "structure" or "pattern" cannot be understood or predicted from the behavior or properties of the component units alone. (Mihata 1997:31)

"Emergent" phenomena, as I use the term, are those that arise from the *interaction* of multiple individual "agents." An emergent phenomenon cannot be fully understood without understanding the properties of the individuals involved, including the rules that govern their behavior. However, such understanding is not sufficient. Understanding the *interactions* of the individuals is also necessary. In short, I am talking about the kinds of systems that are the subject of complexity theory (e.g., Babloyantz 1986; Jantsch 1980; Kauffman 1995; Kohler and Gumerman 2000; Mainzer 1997; 1989; Nicolis and Prirogine 1977). I will make little reference to the details of this body of theory, but two examples of such systems will serve to illustrate the salient aspects of emergence.

In a thin layer of water, the movement of individual molecules is uniform if the temperature of the water is uniform throughout – that is, the movement of the molecules is uniformly disordered. If we begin to heat the bottom of this layer of water, the system becomes unstable, because the denser water near the surface tends to sink while the water near the bottom tends to rise. As the temperature at the bottom continues to rise and heat continues to be dissipated from the upper surface, there comes a point when the uniformity of the disordered movement of the molecules is broken. Convection currents form as warm water rises and cool water descends. These currents are not random but are linked in a pattern of alternately rising and descending currents called Bénard cells (Figure 2.1). (For a more detailed discussion see Nicolis and Prigogine 1989:8-15; Velarde and Normand 1980).

The movement of the water molecules is controlled by relatively simple physical laws that can be understood at the level of the molecule. Understanding what makes an individual molecule behave in the way it does is necessary – but not sufficient – for a complete understanding of Bénard cells. The processes by which these cells form cannot be understood without reference to the interactions of many molecules. Another way of saying this is that, in a system of Bénard cells, the trajectory of one molecule of water is causally linked to that of another molecule with which it has no direct interaction and that may be spatially separated from it by a distance of many convection cells. Both molecules are playing an active part in creating the system and are also controlled by the system. The pattern of Bénard cells is thus an emergent phenomenon that transcends both molecules and whose analysis cannot be reduced to the level of the individual molecule.

Figure 2.1. Viewed in cross section, Bénard cells consist of convection currents moving in opposite directions (after Nicolis and Prigogine 1989, figure 3a). Viewed from the surface (not shown), they form a honeycomb pattern (see Velarde and Normand 1980:92).

Cellular slime mold (*Dictyostelium discoideum*) is an unusual organism that spends part of its life cycle as individual, unicellular amoebas and part of its life cycle as a multicellular organism. The multicellular organism is capable of spatial movement, presumably in search of nutrients. In the course of its life cycle, it becomes differentiated into various kinds of cells and eventually produces spores that germinate into a new generation of individual unicellular amoebas. At each stage, emergent phenomena play a role, but for illustrating the nature of emergence it is sufficient to describe how individual cells aggregate to form a multicellular organism.

In response to lack of nourishment, a few individual amoebas begin to emit a chemical signal called cAMP (cyclic adenosine monophosphate). As the cAMP reaches other amoebas, they respond in two ways. They begin to move up the chemical gradient toward the "pioneer" amoeba. They do so not en masse but in waves of moving and stationary amoebas (Figure 2.2). The reason for the waves lies in the second response:

> An amoeba that is stimulated by cAMP releases it so that the concentration rises and the molecule diffuses into adjacent regions. Amoebas nearby are then stimulated by this diffusing cAMP to produce the signal, which then diffuses and stimulates other amoebas. So the signal propagates across the lawn of cells in a petri dish. But this is not enough to ensure an effective signal: it must also be destroyed; otherwise the whole dish of amoebas would become a sea of cAMP, and no signals would be visible. The amoebas secrete an enzyme, phosphodiesterase, that destroys cAMP. So the substance has a brief life-

time, and the diffusion profile of the signal from a stimulated amoeba
has a steep gradient, generating an effective directional signal that allows other amoebas to use it for chemotaxis (directed movement in
response to a chemical). However, there is a problem here: cAMP released from an amoeba diffuses symmetrically in all directions away
from the source, so amoebas anywhere within the effective range of
the signal could respond. This means that each stimulated amoeba
could become the center of the propagating wave. The result would
be total chaos. This does not happen, as is evident from [Figure 2.2].
The reason is beautifully simple and natural: after an amoeba has released a burst of cAMP, it cannot immediately respond to another
signal and release another burst. It goes into a refractory state during
which it is unresponsive, recovering from the previous stimulus and
returning to its "excitable" condition. Therefore, the wave cannot
travel backward, and the signal travels one way. (Goodwin 1994:50-
51)

Thus, the patterning of the movement of amoebas during aggregation
depends on the chemical responses built into the phenotype of the cell by
its genotype. However, the pattern also depends on the *interactions* of
those cells. It cannot be understood without considering that interaction,
and it is therefore an emergent phenomenon that cannot be reduced to the
understanding of the individual cells alone.

Emergent phenomena can be understood only in terms of the interactions of multiple units – "agents" in the terminology of complexity theory. In both of the preceding examples, the overall pattern of movement
is *created* by the interactions of multiple agents, and in this sense it transcends the individual agent. Emergent phenomena that change through
time also *evolve,* not by natural selection but through the interactions of
individual agents. For example, if we were to revisit the slime mold
amoebas shown in Figure 2.2 at a later time, the waves of movement
would have altered as the amoebas converged on a few centers. The pattern changes because of the interactions of the amoebas. Such emergent
systems are ubiquitous in nature. From snowflakes to hurricanes to the V
formations of flying geese, patterns are created by and evolve through
the interactions of multiple agents.

Much of complexity theory is concerned with systems with very
large numbers of agents whose "rules" of behavior do not change. Bénard cells and the movement of slime mold are examples of such systems. Primate social systems are likewise complex, but they differ in two
ways. The number of individuals in a group is likely to be much smaller,
and each individual can learn from its interactions with other members of
the group. This adds an interesting twist, but it does not change the fact
that primate social systems are emergent phenomena. (The emergence in
this case is at the level of behavior, not of coding.)

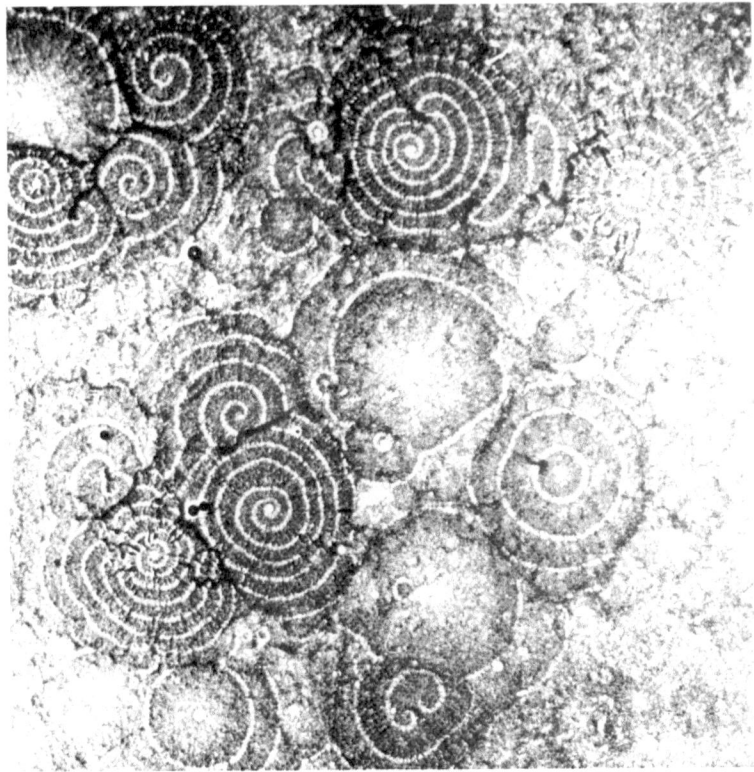

Figure 2.2. Movement of cellular slime mold amoebas during aggregation occurs in concentric waves. Light bands are moving organisms; dark bands, stationary. (Photograph courtesy of Grégoire Nicolis.)

A good example is an account by de Waal (1982) of the activities within a group of captive chimpanzees when the alpha male, Yeroen, was deposed by another male, Luit. This was by no means simply a matter of Luit overpowering Yeroen. Rather, the process took a considerable period of time and involved a third, younger male, Nikkie, as well as the female members of the group. Luit and Nikkie formed a coalition, but Nikkie did not support Luit in his fights with Yeroen. Instead, when Luit and Yeroen were fighting or bluffing, Nikkie confronted the females, who normally would have come to Yeroen's support (and whom Yeroen was often begging for help). Eventually, when Luit supplanted Yeroen, the females ceased to give Yeroen the kind of respect they once had. Nikkie went from having virtually no social standing to second place in the hierarchy. What is more, the trio of Luit, Nikkie, and Yeroen now began to spend more time together and to interact with one another much more than with the females.

In short, the changes in the social configuration of this particular group of chimpanzees involved different sets of interactions among various individuals and groups. Each of these sets of interactions affected other sets of interactions, and the social configuration that emerged was produced by them. It would have been impossible to understand either the process of change or the end result by studying the behavior of individual chimpanzees in isolation. Rather, these could be understood only as emergent phenomena that arose from and in fact consisted of interactions.

Patterns of convection cells or of slime mold signaling are much less diverse and much more monotonous than patterns of primate interactions, even though they involve many more individual agents. Two variables are involved. The first is the complexity of the rules governing the agents' behavior. For water molecules these are the rules of physics; for slime mold, they are chemical and cytological. In a primate society, the "rules" consist of mental coding. The second variable is the extent to which the rules governing the behavior of an individual agent may change as a result of interactions with other agents.

The fact that the rules governing the interactions of water molecules are both few and invariant means that patterns of convection cells differ little from one another except in detail. Primate and human societies are much more variable, because the codes governing individual behavior are more complex, and also because these codes can always change.

This does not mean, however, that social configurations are any less emergent. If we define social configurations as patterns of interaction among individuals, then these configurations are as much products of interaction as are patterns of convection. They too are emergent, and they too transcend the individual. Individual water molecules are active agents in constructing a system of convection cells and at the same time captives of that system. In the same way, individual apes and individual humans are active agents in and captives of the social configurations in which they find themselves.

2.1.3. Socially Constructed, Emergent Coding

I suspect that most scholars who hold culture to be essentially synonymous with social learning or socially transmitted traditions (e.g., Boesch et al. 1994; Boesch and Tomasello 1998; McGrew 1998; Whiten et al. 1999) consider that the differences between human and nonhuman culture are *quantitative*. Our larger brains make it possible for us to learn more complex memes, and language and deliberate teaching make transmission of those memes more efficient. Basically, there is in this view still no *qualitative* difference between human culture and that of other culture-bearing species.

> In an evolutionary progression, if one passes from primates to man the amount and the complexity of the culture increases enormously. If we ask what is different about man that makes this possible, the answer lies in the fact that besides possessing the improved ability to make multiple choice responses and to learn, man has also greatly increased the art of true teaching. One human being cannot only instruct another, but can impart a wealth of information. Furthermore, that information can be transmitted by a powerful language, and it has even been possible to develop ways of writing the language so that communication can take place through the means of artifacts. Finally, because of such storage methods, we have been able to accumulate information. This most recent accomplishment has meant a logarithmic increase in the total stored knowledge that includes all the inventions and innovations of the past. (Bonner 1980:179)

While this is entirely true, it does not cover all that is new in human culture. There is also something qualitatively different – emergent, socially constructed coding.

Among nonhuman species, memes are not emergent phenomena. They do have a certain public character in that they are "shared," but this is analogous to "sharing" the gene for blue eyes with other members of a population. The coding represented by memes is understandable at the level of the individual. An individual interacts with its environment, and on the basis of those interactions either constructs or modifies neural codes that will govern its behavior in the future. It matters little if the relevant part of the environment is the behavior of running water, the behavior of a predator, or the behavior of a conspecific. Each individual constructs coding that it perceives (in terms of its already existing coding) as being beneficial. The codes created in response to this interaction can thus be understood in terms of the individual creating them, and they are therefore not emergent phenomena.

Certainly, the interactions of multiple individuals whose behavior is governed in part by memetic coding will produce emergent social phenomena at the *behavioral* level. The example given earlier of the struggle involving Yeroen, Luit, and Nikkie is a good illustration of this point. However, these emergent phenomena arise in the domain of behavior, not that of coding. The agents in the emergent system are individual animals. Their behavioral interactions produce emergent social systems or social configurations that cannot be understood without analysis at the level of social interaction, but the coding governing each agent's behavior can still be adequately understood at the individual level.

This does not mean that nonhuman social systems cannot be very complex, or that the individuals in such systems are not behaving according to complex, sophisticated, and highly flexible coding. Primate ethology has provided abundant evidence to the contrary (e.g., Byrne and Whiten 1988; Chapais 1995; de Waal 1982; 1989; Dunbar 1988; Goodall

1986; Hinde 1983; McGrew et al. 1996; Quiatt and Itani 1994; Quiatt and Reynolds 1993; Smuts et al. 1986; Tomasello and Call 1994). It simply means that the coding involved is not emergent.

Humans, on the other hand, are governed (in part) by coding that cannot be understood at the individual level alone.[*] It is easiest to grasp this fact by considering codes that are both based on arbitrary convention and serve to coordinate the behaviors of multiple individuals. Take, for example, the red, yellow, and green lights at a highway intersection. These represent an arbitrary convention that facilitates the safe flow of traffic by coordinating the behavior of all the drivers who approach the intersection. While a driver may understand the benefit of traffic lights for himself or herself, this benefit exists only if the convention is "agreed" to by *all* drivers. In the absence of such agreement, the individual's best strategy at an intersection is not adherence to a convention but a combination of caution and bluff.

The latter strategy resembles the monkey Imo's throwing wheat into the water to separate it from sand, because it will work for the individual regardless of whether or not other individuals are guided by it. By contrast, even if there are traffic lights at an intersection, the convention on which they are based will work only if everyone understands and accepts it. Thus wheat washing can be created and understood at the individual level; conventions for traffic signals can be created and understood only at the emergent level.

Examples of indubitably emergent socially constructed coding abound in human life. A chess game, for example, can exist only if the concept of the game, the definitions of the pieces, and the rules of play are agreed on by at least two individuals. One player alone is insufficient. Exogamous clans can organize a society only if everyone agrees on the definition of a clan, the definition of marriage, and the rule of exogamy. If only one person adheres to the concept of exogamous clans, society will be organized along other lines in spite of him or her.

Among the most important of emergent codes are the semantic and syntactic conventions that make up languages. Unless everyone in a conversation uses the same conventions, linguistic communication will not exist. If one wants to talk to another English speaker, one has no choice but to use English words and English conventions for indicating tense, number, and so forth. It is possible for one individual to make up his or her own language, but no communication will take place unless at least one other person adheres to the same linguistic coding.

---

* Eve, Horsefall, and Lee (1997:36) have also argued that culture is an emergent phenomenon and that the individual and emergent levels can be neither separated nor reduced one to the other. However, their concept of culture is rather different from mine.

Coding can become emergent only if it is created and maintained or modified through social interaction among multiple individuals. For example, until 1967, everyone in Sweden drove on the left side of the road, by legally binding convention. In 1967, the Swedish government decreed that, as of a given date, everyone would instead drive on the right. Thus a new convention was created that governed the behavior of all Swedish drivers. This convention was created by Swedish society – by the interactions of Swedish administrative, political, and legal institutions and Swedish voters – and it worked because it was accepted by Swedish drivers. For this reason, the new convention was emergent in nature.

This does not mean that an individual cannot create a code that becomes emergent. For example, I have acted individually in creating an idiosyncratic definition of the word "culture." As I sit at my desk writing this paragraph, the definition has not been adopted by anyone else and so does not constitute emergent coding. I hope that by the time you, the reader, reach this paragraph, you will have understood and adopted the definition for use within the limited context of this book. If you have, then by that act you have turned an idiosyncratic code into an emergent code. It is not necessary that you agree with me or with my analysis of culture for this to be the case. All that is necessary is that when you read my word you take it to mean what I meant when I wrote it. If so, then communication exists, because we share an emergent code. It is this social interaction between me as writer and you as reader that gives the code its emergent nature.

In certain cases, one individual has the power, for whatever reason, to impose idiosyncratic codes on others, so that they govern everyone's behavior. For example, during the 1980s, my parents' mailing address was changed. A Postal Service employee in central Oregon had decided that it was more logical to number rural postal boxes according to a map grid system than sequentially along a delivery route. This produced very long numbers that were hard to remember and that few people liked. Nevertheless, anyone who wanted mail delivered to the right place was obliged to use the new system. That it was imposed by one bureaucrat, without consulting postal customers, did not make the new system any less emergent. What made it an emergent coding system was not people's motive for adopting it but rather the fact that it worked because everyone adopted it. Residents informed their correspondents of the new numbers, and when new mail arrived, postal workers knew where to deliver it.

To avoid confusion, I must pause here to clarify my terminology. The reader may wonder what distinction I make between "socially created coding" and "emergent coding." The answer is none. The difference is one of emphasis only. To avoid introducing new jargon, I will generally use the former rather than the latter term. However, either term should be read to mean "coding that is created and maintained or modified through social interaction among individuals and that is therefore intrinsically emergent."

Note that the "creation" involved is the creation of a code that transcends the individual. This is necessarily a social, not an individual act. In the examples just given, one individual created an idiosyncratic code. The adoption of this code by others constituted its social creation and moved the code from the idiosyncratic to the emergent level. (By contrast, the adoption of a meme does not move the meme from the individual to the emergent level. There is no social creation involved, just social transmission.)

It may further clarify the difference between emergent and memetic codes if we consider (1) the consequences to the individual of rejecting each of them and (2) what the individual must do in order to change each of them.

If an individual either fails to learn or simply rejects a meme, the consequences depend on the nature and value of the meme. A young mouse from the example cited earlier who fails to learn that the smell of a rat indicates danger may well pay with its life. The macaque who does not adopt sweet-potato washing will eat gritty potatoes, a matter of much less import. In either case, the consequences come directly from the environment as a result of the way the individual deals with that environment.

If an individual fails to learn or opts out of an emergent code or coding system, there are three classes of consequences. The first is analogous to what happens if one fails to adopt a meme. If, for example, a stubborn Swedish farmer had refused to drive on the right side of the highway, he likely would have paid with his life. This example differs from those of the mouse and the macaque only because the environment involved is not the natural environment but the behavior of conspecifics. This is not a significant difference – the behavior of conspecifics is still a part of any organism's environment. For example, among vervet monkeys,

> in a typical interaction involving two playing infants, one or both of the infants will scream when play becomes rough, and both mothers will come running. The dominant mother will then threaten or supplant the subordinate mother and her infant, and the subordinate pair will retreat. ... from a very early age group members behave differently toward the infants of high- and low-ranking mothers. High-ranking infants are often more sought after as play and grooming partners, and in many other ways interactions with them are carried on in a more careful manner than are interactions with infants of lower rank (Lee 1983; Nicholson 1987; Whitten 1982). ... In rhesus

macaques (Datta 1983), juveniles consistently challenge adults who
rank below their mothers but rarely challenge adults who rank above
their mothers. This suggests that a juvenile monkey learns about her
"expected" dominance relations with others at a very early age. She
seems to do so both through her own experiences and by observing
interactions between her mother and other group members (Altman
1980; Berman 1980; Datta 1983; Horrocks and Hunte 1983). (Cheney
and Seyfarth 1990:31)

In other words, one of the things any primate must learn is how other
individuals are likely to react under given circumstances. The reason, of
course, is that the behavior of other individuals will have an effect on
one's own life. In all primates, not just humans, this ability to observe,
predict, and adjust one's behavior to social facts is highly developed,
with the result that primate social systems tend to be both complex and
flexible.

The penalties for failing to predict the behavior of other individuals
come from the behavior of conspecifics. The young macaque who fails to
recognize that his mother is subordinate to his playmate's mother risks a
painful lesson if he is too rough with that playmate, just as the mouse
risks being eaten by a rat because it fails to learn that rats are dangerous.
In this respect, the death of a stubborn Swedish farmer who refuses to
accept that everyone else is driving on the right side of the road is no different just because it stems from a refusal to accept an emergent code
rather than from an inability to learn, as an individual, about the behavior
of others.

A second kind of consequence faced by an individual who fails to
accept an emergent code or system of codes is simply that he or she is
left out of the social system or social activity that the code produces. This
may be of little consequence. For example, I personally do not feel
handicapped because I never learned the rules of bridge. However, because I have not done so, I cannot join in a game. In other cases, the consequences may be more severe. For example, in the unlikely event that
someone in a hunter-gatherer band refused to learn the conventions controlling communal hunts, he would be unable to participate in those hunts
and might be denied a share of the prey.

This kind of exclusion is not the same thing as not learning how to
deal with others socially. All social mammals, whatever their individual
social skills, are nevertheless involved in social interactions. Being socially inept means failing to accomplish one's goals in a social setting,
whether these have to do with rank, access to food, access to mates, or
something else. If one does not learn to play bridge, the consequence is
not that one fails but that one cannot even play the game.

In many cases there may be a third kind of consequence. An emergent coding system may include the requirement that all individuals ac-

cept and adhere to that system, and that those who fail to do so be punished. This is typical of some religious systems, many moral codes, and of virtually all legal systems. In such cases, if one fails to accept and adhere to a code, one will be punished by other members of society. The punishment will be prompted by the same set of codes that one has rejected. Sometimes the punishment is harsh, including torture or death. Sometimes it is limited to mild ostracism or simply the withholding of social approval, as when someone wears a necktie that is unfashionably narrow or eats his salad with the wrong fork. It is true that even in the absence of emergent coding, individuals will still use coercion to enforce their own interests or those of relatives or allies. Yet among humans, individuals often use coercion to enforce an emergent cultural code, regardless of their own individual interests.

If an individual animal is dissatisfied with a meme, it is free to change it. For example, the tits I described earlier all opened milk bottles, but they opened them in different ways. Moreover, different birds apparently preferred milk bottles with different colored caps (Fisher and Hinde 1949). Modifying a meme is done in accordance with one's own internal codes and one's own experiences. An individual has no such freedom with regard to an emergent code. An emergent code is not emergent until it is accepted by more than one individual. Therefore, in order to either create or modify an emergent code, one must somehow influence others to adopt it. In some cases, such as that of the Oregon postmaster, one individual may have the power to impose his or her will on others. More often, the process involves persuasion, negotiation, and compromise. In all but very small social groups, even a tyrant depends on the loyalty and support of his subordinates to impose his will on others.

Thus, from the laws that govern a nation to the rules of a children's game, emergent coding is usually the result of a more or less complex process of coercion, negotiation, persuasion, and compromise, a process that involves at least a portion of those affected by the outcome. This does not mean that everyone is equally influential in the process, but simply that the process involves more than one person. One person may invent a new game, but the game will not exist as a game unless at least one other person is persuaded to learn its rules.

The emergent nature of cultural coding is the central concept in this book. As will be seen in chapter 4, such coding appears to be unique to humans. This is not to say that no emergent phenomena are to be found among other species. It is becoming apparent that, because different genes interact with one another, the genotypes of all species are characterized by emergent phenomena (Kaufman 1993). The same is true of the phenotype, where different parts of the body interact with one another and where ontogenetic development is characterized by emergent phenomena (Goodwin 1994). Networks of interacting neurons make the functioning of the brain an emergent phenomenon. Above all, social interactions, being interactions among individuals, produce emergent phenomena at the *behavioral* level, patterns of social behavior that cannot be understood without investigating those interactions. What seems to be unique to humans is the emergent nature of a significant portion of the codes that exist in our minds or brains and that influence our behavior.

Socially constructed, emergent coding makes the human way of life different from that of all other animals. It lies at the very core of human culture. There is, however, much more to human culture than just socially constructed coding per se.

2.2. SOCIALLY CREATED CODING AND HUMAN CULTURE

For a large set of Holocene humans – those people living today and those for whom we have reasonably good historical or ethnographic records – we can be confident that we understand the general characteristics of their cultures. It is clear that for all of them (whether the people are hunter-gatherers, agriculturalists, or members of urbanized, industrial societies),

- Culture is based on socially created codes.
- Socially created coding provides motivation for the individual's behavior.
- An all-inclusive system of emergent coding pervades and absorbs into itself almost all other coding and almost everything else perceived or thought of by humans.

Yet there is no a priori reason that socially created coding could not at one time have existed without the two other characteristics of Holocene culture. Socially created coding can, at least theoretically, be a very simple phenomenon. This may not be obvious from the examples given earlier in this chapter, most of which were drawn from modern contexts. An imaginary heuristic example might be clearer.

Wolves hunt large game such as moose (*Alces alces*) cooperatively. Whereas it would be difficult for a single wolf to kill such a large animal, a group can tire its prey by taking turns pursuing it and can kill the victim by mobbing it. Since a moose is large, it yields enough meat for many hunters. As a result, it makes sense to hunt cooperatively as a pack.

Imagine a group of early humans setting out to hunt large game with relatively unsophisticated weapons. For them as for the wolves, it would pay to hunt as a group. If they were capable of agreeing beforehand on a modus operandi, their chances of success would be even greater. For example, they might agree on a strategy for driving a herd of bison or other large bovids over a cliff, so that each individual would know where he or she should be and what he or she should do. This would certainly be a form of socially constructed, emergent coding. Let us suppose, however, that for our imaginary group the creation and use of socially constructed coding stopped there, and that in all other respects their behavior was governed only by the kinds of codes characteristic of wolves and chimpanzees. Clearly, their way of life would be very different from that of living humans.

2.2.1. Motivation and Susceptibility to Socially Created Coding

First, the *motives* of each individual would be no different from those of a wolf in a cooperative hunt. The codes that stimulated him or her to hunt as part of a group would not be emergent. The motives would be hunger, on the one hand, and a calculation (conscious or otherwise) that cooperation would be advantageous for satisfying that hunger. In our imaginary group, the individual's behavior during the hunt would be governed by the emergent codes constituting the agreed-upon strategy, but the *motivation* would be of an entirely private, individual nature. The socially created coding would provide instructions for how to hunt cooperatively but not a reason for doing so.

In all known present-day human cultures, socially created coding also seems to provide motivation for behavior. People are moved to remain celibate by the hope of eternal life in heaven, to die in a suicide mission by the desire to serve their emperor or God, or to toil in low-paying jobs by dreams of academic glory.

This raises an interesting problem from an evolutionary perspective. Because emergent codes are created through the interactions of multiple individuals, there is no a priori guarantee that they will produce behavior that will benefit any given individual. If natural selection acts on the individual, it follows that it should quickly destroy any tendency to obey codes that might reduce the individual's evolutionary fitness.

When codes are generated externally – by multiple individuals – no one person can be assured that the results will be beneficial to him or herself, or even that they will not be downright deleterious. In addition, it is a characteristic of complex (i.e., emergent) systems that their evolution is unpredictable. As a result, whenever multiple individuals interact to create coding, it is always possible that the system will produce unintended consequences, trapping individuals in a system of coding that benefits no one.

All organisms are parts of systems (ecosystems, social systems, etc.) that may threaten their individual evolutionary success. Such systems constitute environments in which the individual competes. Natural selection tends to produce (private) codes that let individual organisms inter-act with these environments in ways that maximize their chances of re-productive success. These could include a code for accepting socially created codes such as procedures for carrying out a game drive – as long as following those codes contributed to the individual's fitness. This is exactly what our imaginary group of hominids is doing. However, as soon as individuals permit emergent codes to motivate their behavior, they run the risk of permitting those codes to cause them to behave in ways that lessen their individual evolutionary fitness. The codes are no longer just part of the environment, but part of their coding for dealing with the environment.

The question therefore arises, how could natural selection have failed to prevent the evolution of a willingness to let socially created (and therefore external) codes motivate one's behavior? The same question arises with regard to the apparent propensity of our species to act altruistically, helping others at one's own expense. This is a complex question that is the subject of a large body of literature. As will be seen in the next chapter, nothing about the question is simple –not even the definition of evolutionary fitness. But until we understand when, how, and why cultural coding came to provide the motivation for individual behavior, we will not understand the evolution of human culture and of the human way of life.

2.2.2. Socially Created Coding as All Encompassing

The way of life of our imaginary group of early humans differs from that of recent humans in another fundamental way. In their lives, socially constructed coding is restricted to a narrowly circumscribed activity. Yet among all the humans who are living today, or who are known ethnographically or historically, such coding is pervasive rather than restricted.

Let us give our imaginary group a second set of socially created coding, a simple language. The language consists of phonological, semantic, and syntactic conventions that permit these people to express and to understand ideas about their environment. Thus, their language may have words for concepts such as "berry," "ripe," "three," and "day," as well as syntactic conventions for expressing relations among them, such as, "The berries on the other side of the ridge will be ripe in about three days." Clearly this language constitutes a set of emergent coding, and it would be of great utility in teaching children, organizing cooperative activities, and exchanging useful information.

It is limited in scope, however, in comparison with the language and culture of present-day humans. All it does is to permit communication about things in the natural world or about concepts about those things that would probably exist anyhow. For example, a father could explain to his not-too-bright son that a spear must be sharp if it is to be effective. However, he is only communicating something about the real world that he knows already, without language or symbolism.

We humans, however, use language to construct a large repertoire of "things" that have no existence outside a symbolic cultural context and that depend on that context for their very existence. Such "things" pervade the entire environment in which we as humans live our lives. They come in an almost infinite variety: beings (deities, ghosts), social roles (presidents, bridesmaids), objects (scepters, stop signs), concepts (sin, authority), acts (baptizing, promising), values (virtuous, chic), and so forth.

Still, this tendency to create cultural entities that have no existence in the concrete world around us is not the most important difference between the language or emergent coding of our imaginary group and those of living humans. Whatever its origin, the effect of this tendency is to make possible a much more significant development. Socially constructed codes merge into pervasive, all-encompassing, ubiquitous systems of thought that incorporate almost everything that humans think or perceive. Animals may become totems, assimilated into a framework of kinship and religious belief that is entirely cultural. Natural relationships such as motherhood or siblinghood are given cultural meaning beyond their biological meanings. Indeed, kinship is usually defined culturally, with cultural definitions taking precedence over biological ones.

Even private codes are incorporated into emergent coding. Emotions such as anger, sexual desire, and fear are given cultural meanings and cultural values that depend upon the culturally defined contexts in which they occur. In short, almost everything a person does, thinks, or feels comes to have cultural meaning.

It is not that socially constructed coding displaces or replaces either the natural environment or individual or memetic coding. Rather, it assigns them cultural meanings and values and uses them as cultural symbols. The result is that while the behavior of an enculturated individual is still guided in part by individual and memetic coding, everything he or she does, feels, or thinks is now enmeshed in a cultural system. Although it is possible and even desirable to distinguish analytically between the natural and the cultural environments, or between individual and emergent coding, in practice the enculturated individual can never ignore the cultural meanings of natural phenomena or the cultural meanings and consequences of behavior guided by individual coding.

By contrast, the members of our imaginary group of early humans, in common with wolves and chimpanzees, use private coding to perceive and interact with their environment. When they do make use of socially constructed codes, it is in the limited context of specific social activities that are coordinated by simple emergent codes.

I am certainly not the only person to argue that culture provides an all-pervading matrix of meaning to human experience and human behavior (e.g., Durkheim 1915 [1965]; Geertz 1973:5; Rappaport 1999:8-9). These authors' theoretical perspectives on culture differ from one another's and from mine, but I follow in their footsteps by seeing culture as an all-encompassing intellectual or ideational environment.

2.2.3. Memetics in the Context of Human Culture

Human culture, then, is based on a form of coding that at some point in the course of our evolution was added to already existing forms of coding, whether genetically determined, learned, or memetic. Thus, in principle, present-day human culture includes a memetic element. However, the emergent and all-encompassing aspects of human culture affect memetics in three ways:

- First, in the presence of language, most memes will be codes, not behavior.
- Second (and this is much more important), when socially created codes guide and motivate individual behavior, and when all things (including memes) come to have cultural meaning and positive or negative cultural value, memetic selection can no longer be assumed to take place at the level of the individual.
- Third, a new entity, the culture trait, comes into being. As I use the term, a culture trait is in some ways analogous to a meme, but it consists of coding created socially and transmitted from group to group rather than from individual to individual.

With the advent of language, there can be little question that some memes will belong to the domain of coding. Earlier in this chapter I mentioned that among nonhuman species, what was transmitted from one individual to the next was usually not a code but a behavior. When one individual explicitly teaches another, however, what is transmitted is information about how to do something, when to do it, or something similar. Such information is coding, not behavior. I discuss in a later chapter claims that such deliberate teaching takes place, without language, among apes.

In the case of humans, it would be difficult to believe that, in the presence of language, coding was not transmitted from one individual to another. As a medium, language transmits information – that is, coding – rather than behavior. The very act of creating or learning a language involves a mindset based on shared coding. Thus, in the context of emergent culture, many but not all memes will consist of coding rather than behavior. For example, there are many ways of making chili con carne, and it may be that in some cases one person learns how to make it by watching another person. When a recipe for chili is written down in a cookbook, or when someone explains verbally how to make his or her version, then the recipe or explanation is clearly a form of coding.

Human culture has another, much more significant efffect on the way ideas are transmitted socially. It is the independent selection of memes by individuals that makes memetic evolution an essentially Darwinian process. In the context of human culture, this independence is seriously compromised. The reason is that memes, like virtually everything else humans either do or pay attention to, are caught up in the web of culture in the sense that they are assigned cultural meanings, values, and so forth that transcend the private coding of the individual. When an individual either adopts or rejects a meme, he or she is also, like it or not, performing a cultural act. As a result, the individual's decision is guided not only by internal private coding but also by external cultural coding.

For example, when a chimpanzee makes a wand to fish for termites, she needs to consider whether it is stiff enough, flexible enough, smooth enough, and of the right size to do the job. Unlike humans, she need not worry about whether the particular form or color of the wand is too flamboyant, too passé, too masculine, and so on. Thus she can make a wand on the basis only of her own internal coding, which informs her of how well it will function physically. If she were human, she would also need to worry about how well it would function socially, in terms of socially created cultural coding.

This is not to say that ideas and behaviors do not spread among humans as memes spread among nonhumans. However, the process virtually always involves not only private decisions but cultural factors as well. Let me give two examples from modern life. Someone, somewhere, invented the pocket protector, a plastic insert that protects the breast pocket of a shirt from ink leaking from pens. Although it does this job well, it is almost never seen today, at least in North America, because it has also become the stereotypical symbol of the nerd, of someone with an enthusiasm for unfashionable activities such as engineering combined with an embarrassing lack of social graces. Of course, there is nothing inherent in the pocket protector that gives it this meaning. The meaning is assigned by culture, but it has nevertheless played a major role in selecting against the adoption – by individuals – of pocket protectors.

Another example is provided by fast-food hamburger restaurants. These are very popular in the United States and are becoming popular in France as well. Interestingly, the cultural meaning of fast-food restaurants is quite different in the two countries, and as a result the customers in the two countries differ.

In the United States, the most common positive response to fast-food restaurants is that they are a quick, inexpensive, and convenient way of getting a meal. The most frequent negative response is that the food tends to contain unhealthy levels of fat, cholesterol, and sugar. The clientele at fast-food restaurants includes people of all ages, as individuals, in families, and in groups of unrelated people. In France, on the other hand, there is a sharp age division in attitudes toward fast food. For teenagers and young people it is a way of identifying with popular culture and with one's age group, and a way of distinguishing oneself from the older generation. For older people, it is a threat to French culture, to traditional norms and attitudes concerning meals and eating. The result is that older people and families frequent fast-food restaurants much less often than they do in the United States.* In this case, emergent cultural attitudes affect which segments of the population have adopted the fast-food meme.

This example illustrates the problem of analyzing the evolution of human culture in a situation where socially created as well as private coding determines the adoption or rejection of a meme. It also illustrates the third change brought about by the appearance of culture. A new phenomenon arises in the context of socially created culture, one that resembles a meme but that actually belongs to the emergent rather than the individual level. This is what has traditionally been referred as a "culture trait" or by some similar term. Culture traits are units or complexes of socially created coding that originate in one society and then are adopted by other societies or subsets of other societies.

For example, a religious movement arose among the native tribes of the western United States in the late nineteenth century, tribes undergoing severe stress as a result of white conquest. The movement foresaw a return of the dead ancestors, a renewal of game now become scarce, and a general return to better times. This revival was to be hastened by the performance of a Ghost Dance, which gave the movement its name. The Ghost Dance had its roots in the Prophet Dance of the tribes of the Northwest (Spier 1935). In 1870 and again in 1888, the Ghost Dance was stimulated by prophets of the Paviotsos in northern Nevada. From there it spread to the tribes of the Great Plains, where it became one cause of warfare against whites (Mooney 1896).

It is instructive that among the Lakota, an emphasis on fighting against and destroying or driving out whites became a part of what, among the Paviotsos, had been an essentially peaceful if not pacifist

---

* I admit that this discussion of fast-food restaurants is based on my own observations, which would hardly meet scientific ethnographic standards. My purpose is simply to illustrate a point, not to provide a comparative ethnography of French and American eating habits.

movement (Lowie 1954:181). It is equally significant that in spite of its appeal to a large number of tribes, the Ghost Dance was rejected by the Navajos, who for religious reasons considered the dead to be extremely dangerous and who looked upon the possibility of their return with serious misgivings (Hill 1944).

Thus, the processes by which culture traits spread (or fail to spread) among groups are equivalent to the processes by which memes spread among individuals. Culture traits are adopted when they are perceived as potentially beneficial and rejected when they clash with existing codes. However, they belong to the emergent level of society and culture, not to the individual level where memes reside. Like all emergent codes, they are created by interactions among multiple individuals, and they can be adopted only by groups of individuals.

The emergent nature of human culture poses a challenge for memetic analysis. Some very meme-like phenomena are indeed present in human culture. Moreover, these meme-like entities are in fact transmitted from one individual to another. The problem is that what makes a Darwinian analysis of memetic evolution possible is selection at the level of the individual. In the context of human culture, this aspect of memetics breaks down.

The meme is also an inherent part of the cultural system, so it becomes in a very real way an emergent as well as a memetic code. Its evolution is determined (at least in part) by the processes that drive the evolution of emergent or complex systems. This mechanism does not involve selection but consists of interactions among multiple agents.

This does not mean that adherents of memetic or dual inheritance models of human culture have it all wrong. There is a strong component of this kind of transmission in human culture. However, the analysis of even the memetic component of human culture does need to be altered to accommodate and account for, in some way, the emergent aspects of human culture.

2.3. FURTHER OBSERVATIONS ON HUMAN CULTURE

To this point, I have presented only a definition of emergent coding and human culture. The resulting picture is so stripped down that it risks being misunderstood. It would be a good idea, therefore, to flesh it out by discussing a few additional points concerning the way I conceive of human culture.

Socially constructed coding, by its nature, arises from the interactions of more than one individual. However, the ways in which such codes are created are varied. At one end of the continuum, they may be created cooperatively. At the opposite extreme, they may be imposed by force. Coercion as a source of emergent coding cannot be eliminated.

I gave an example earlier of one form of "coercion," the imposition of an unpopular postal box numbering system by a Postal Service bureaucrat. In that case, the creation of the numbering system could be attributed to a single person. Mail delivery, however, depended on the use of that system by people who addressed mail, sorted mail, and delivered mail. Thus the emergent nature of that system of addressing mail was unaffected by the fact that it was its having been imposed by fiat. In addition, the power of the bureaucrat in question derived from a whole set of cultural coding (the definition of the United States Postal Service, its rules and regulations, its personnel structure and table of organization, etc.).

Yet we can easily imagine a very small society in which social codes for, let us say, the coordination of foraging activities is controlled by one individual, X, through the threat or use of purely physical force. Assuming that the codes in question can be understood only in terms of the interactions of multiple individuals (e.g., X lies in ambush while Y and Z are ordered to drive game toward him), then the coding is emergent, regardless of the fact that Y and Z have no desire at all to hunt with X. Nor should this situation be confused with, for example, the forceful stealing of food by a dominant individual. In the latter case, only behavior is involved. In the former, it is coding that is imposed by force.

Examples of cultural coding that is accepted involuntarily are easy to find. In both the southern United States until the 1960s and in South Africa until the 1990s, black citizens were forced to live by oppressive rules that they did not accept voluntarily and for which they saw no moral justification. In spite of their unwillingness, most by necessity obeyed the rules of segregation and apartheid. For example, they understood and obeyed rules about where to sit on a bus or which drinking fountains to use. It is because these rules governed their behavior, and not because they were accepted willingly, that I would include them within my definition of socially constructed, emergent cultural codes.

Many cultural codes, of course, are not imposed by pure brute force. Rather, they are the products of less one-sided processes, such as discussion and voluntary agreement. The process may consist of negotiation or of political deal-making, in which each individual aims for a result that is as close as possible to what he or she would like.

This raises the next point that I want to make about human culture. In the process of social interaction, each individual acts according to his or her own coding. This includes genetically determined coding, individually learned codes, memetic codes, and emergent, cultural codes.

These codes at times reinforce each other, but at other times they conflict and compete with one another for control of the individual's behavior.

Because genetically determined codes such as hunger, sexual desire, and fear are so fundamental and so strong, they have a very important influence on the individual's behavior. To a great extent, individually learned coding and memetic coding are constructed by the individual because they are perceived as satisfying genetically determined coding. As a result, each individual to a large extent acts in terms of his or her evolutionary self-interest, even in the context of human culture.

When cultural coding requires behavior that is perceived as not being in the interest of an individual, that individual is likely to try to change the cultural coding. Because other individuals are doing the same, and because cultural coding is created by the interactions of multiple individuals, it is almost inevitable that not all individuals can be successful in this attempt.

> Culture would represent the cumulative effects of inclusive-fitness-maximizing behavior ... of the entire collective of all humans who have lived. ... If this theory is appropriate, then aspects of culture would be expected to be adversary to some of the wishes of each of us; few aspects of it would be viewed with equal good humor by all of us. (Alexander 1979:68)

Culture is far from being a static, immutable force that rigidly determines the behavior of all members of a society. Rather, it is a dynamic phenomenon, of which individuals are at the same time both the creators and the captives, and which is also only one of the factors determining their behavior.

It is true that we as individuals either voluntarily or involuntarily accept the dictates of cultural coding even when they conflict with our individual, internal coding. However, this trait also gives the individual a new weapon in the competition with others. Culture becomes a way of manipulating the behavior of other individuals. Cultural codes (rules, values, etc.) can variously be invoked, manipulated, or altered in order to influence their behavior. The result is that culture is at one and the same time an arena in which the struggle for individual success is, in part, played out, an object of competition, and a means of competing. Marx's (1884 [1970]) famous characterization of religion as the "opium of the people" reflects this dual propensity to submit to culture and to use culture to manipulate the behavior of others.

To give an example: if one wants a playground for one's own children, but wants it built at someone else's expense, one might do two things, both of which involve changes in cultural coding. One might try to manipulate the political system to have the local government fund a new playground in one's neighborhood. One might also try to change the tax code to shift the burden to someone else. Thus cultural coding becomes a weapon in social competition. It can also become a weapon in the struggle to determine what the culture will be. In the United States, liberals and conservatives, Democrats and Republicans all invoke cultural icons such as the "founding fathers" or the "framers of the Constitution" not only to convince voters to elect them but to convince voters to support or oppose changes to the laws governing the country.

Finally, it should be noted that not all individuals will have the same set of cultural codes (Chase 2001b; Hutchins 1995; Netting 1974). There are several reasons for this. Many such codes are specific to particular activities or to particular groups of people. Males and females, for example, may each possess specialized cultural knowledge not shared with the other sex. Initiates or members of secret societies may have closely guarded cultural codes. Specialists have cultural expertise that others do not. In some societies, only scribes can read and write. Mathematicians, lawyers, and astrologers have mastered subsets of culture that others are less familiar with.

However, there is another important reason why different individuals possess different cultural codes. In the segregated South, everyone understood that blacks were required to sit at the back of a bus, behind the white passengers. Both blacks and whites shared this cultural code. There was no such agreement, however, on the cultural coding used to justify this rule. Southern whites (for the most part) had one set of beliefs and values concerning the nature and meaning of race, which was most emphatically not shared by blacks, who had a whole different set of values and beliefs. For decades, blacks were coerced into obeying the laws and rules of segregation even though they did not accept their moral authority.

Eventually, partly through civil disobedience and partly by using shared cultural beliefs and values to sway public opinion, they succeeded in getting the laws changed. The laws and all the conflicting beliefs and values were parts of the culture of the United States; all arose from the interactions of multiple individuals. The dynamic nature of culture is reflected in the fact that one group of people within the society used one subset of the culture to bring about change in another part of the culture, a change that was advantageous to them.

Finally, there is room for some more or less random variability in cultural coding. Emergent coding will work as long as every individual's version produces behavior that is consistent with the "purpose" of the coding. Language provides an excellent example. Individual speakers of a language and speakers of differing dialects have somewhat variant phonemic coding. That is, they tend to pronounce words somewhat differently. A considerable degree of such variation is possible before people fail to understand each other.

As another example, the Commonwealth of Pennsylvania has on its books a very specific regulation about how far one must park from a fire hydrant. I learned this rule when I took my written driver's test and promptly forgot it. Yet I have never been ticketed for violating the rule – my own idiosyncratic idea about how much space I must leave is sufficiently close to that of the police to avoid punishment. The same is true of millions of other Pennsylvanians who have forgotten the official distance and whose idiosyncratic ideas of what is appropriate undoubtedly differ from mine.

In short, an emergent culture is not a static, monolithic entity shared equally by everyone and rigidly controlling everyone's behavior. It is a varied conglomeration of different sets of coding produced by the interactions of different groups of individuals for different reasons, sometimes very specific and sometimes very general, sometimes widely shared and sometimes contentious, sometimes in a stable state but always potentially subject to change.

Nor, for that matter, are the boundaries of a culture clear-cut. If one group of people were completely isolated from all others, then their culture would have clear-cut boundaries. But in a world where one group of people interacts with other groups, those interactions will produce emergent codes. These may vary in scope and complexity from simple arrangements for periodic trading between tribes of hunter-gatherers to the international air traffic control system, but they nevertheless are shared by different peoples. One set of cultural codes (a religion, a recipe, a decorative motif) may be shared by peoples who, on the whole, have quite different bodies of culture. Pilots and air traffic controllers around the world share a body of socially created coding with one another that is not shared with most of their compatriots. Yet in other respects, the cultures of pilots or air traffic controllers from various nations are quite different.

Isolating a single culture, then, is much like isolating a weather system. It is more or less an arbitrary matter. There may be a storm on the east coast and a high pressure system on the west coast, but they both share the influence of the jet stream. By the same token, French culture may differ in many ways from Mexican culture, but they share Catholicism.

2.3.1. Culture as Superorganic

The idea that culture is in some way a "superorganic" phenomenon that transcends the individual is an old one in the history of anthropology and sociology (e.g., Durkheim 1938 [1964]:xlvii-lvi, 1-13; Hanson 2004;

Kroeber 1917; White 1949; 1975). This raises a somewhat difficult issue, in that the only real locus of culture is the individual. In other words, the only physical reality that culture has is in the neural structures of individuals' brains. Some scholars have dealt with this apparent problem by arguing that culture represents an abstraction – a sum, average, or other summary – of individual codes, or else the distributions of codes over various individuals in a population (e.g., Atran 2003; Hannerz 1992; Rodseth 1998; Sapir 1938; Schwartz 1978). My notion of culture sides with the first group of theories insofar as I see culture as in some way transcending the individual – but with some significant differences.

It is beyond the scope of this book to address these issues in any detail, but they merit a brief comment. The notion of culture as an abstract summary or a distribution of individual coding is applicable to the memetic model of culture, but it misses the point of human culture as an emergent phenomenon. Human culture shares with all other emergent phenomena the strange fact that in one sense it is based on the level of the individual agent and yet at the same time it transcends that level. Thus the emergent pattern of Bénard cells consists of individual water molecules and their movements, yet it transcends the individual water molecule. It is neither the sum nor the average of their movements. In fact, both the sum and the average of the movements of all the molecules equal zero, since the upward movement of rising molecules is offset by the downward movement of sinking molecules, and horizontal movements are likewise balanced.

Rather, the emergent phenomenon consists of the patterning of movement. Movement belongs only to individual molecules, but the patterning of these movements is as real as the movements themselves, and this patterning arises from the interactions of those same individual molecules. Moreover, that patterning is not just an empirical distribution of movements, but a system that both results from and determines those movements.

By the same token, neural codes exist only in the brains of individuals. Nevertheless, they have a patterning that is just as real as the neural structures themselves, and one that arises from the interactions of the individuals involved. Thus, for both convection cells and individual molecules and for emergent codes and individual neural structures, the two phenomenal levels are as closely linked as the two sides of a coin. There is no inconsistency in seeing both as equally real (pace O'Meara 1997).

2.3.2. Complexity Theory and Culture

There is, however, one way in which the emergent patterns of cultural coding differ from those of convection cells, waves of moving slime mold, and most of the other phenomena studied by complexity theorists. In addition to the dichotomy between the levels of individual agent and emergent phenomenon, culture also involves a dichotomy between coding and behavior. Two kinds of feedback are involved. In classical complex systems, feedback exists between the behavior of one agent and the behavior of other agents. This is true for humans as well, in the behavioral (social) domain. There is also a feedback between coding and behavior. Social interaction produces coding that in turn affects the social behavior that then acts upon the coding. Cultural codes influence individuals' behavior in social interactions. One product of social behavior is the creation and maintenance or alteration of those same cultural codes.

Many applications of complexity theory to human behavior deliberately bypass the emergent level of cultural coding. They consist of models that use very simple, invariant codes to account for complex patterns of behavior. A good example is an intriguing study of the Balinese system of subaks, or rice-farming cooperatives (Lansing 2000). Briefly stated, subaks are rice-farming cooperatives associated with tertiary-level irrigation systems that cover about 50 hectares and include about 100 farmers. A complex pattern of cooperation exists among subaks. Damage from insect pests is reduced as the number of adjacent subaks that plant rice simultaneously is increased. Yet as more adjacent subaks plant simultaneously, they face greater problems with water shortages at early stages in the crop cycle. Many patterns of simultaneous or nonsimultaneous planting are possible, and finding an advantageous one is difficult, yet somehow the Balinese farmers succeed.

Lansing found by computer simulation that if each subak simply imitated the immediately adjacent subak with the best yield, eventually a pattern of cooperation would emerge that closely resembled the pattern on the ground. As this particular pattern developed, yields would increase until all subaks had reached a local optimum. Thus, a very simple rule that does not change as a result of interactions among subaks can account for the complex pattern of simultaneous and nonsimultaneous planting found on the ground. Note that the only feedback in this system (once the rule has been agreed upon in the beginning) is in terms of the behaviors of the various subaks.

Such applications of complexity theory to human or animal behavior consider feedback at the behavioral level, with the coding involved held constant. Research is also being done, particularly in economics, into complex systems in which the behavior of the agents changes in response to the state of the system as a whole (eg., Arthur 1999). To illustrate how such an analysis might work, a change in the characteristics or behavior of one species in an ecological community may lead to evolutionary changes in the behavior of other species that interact with it. Because the "rules" governing their behavior changes, one must usually conclude that the coding of individual agents is changing. But even analyses such as these bypass the level of emergent coding. It is the individual coding of the agents that is considered to change, not emergent, socially constructed coding. (In the ecological community example, it is the genetic coding of the individual species.)

I am in no way challenging either the validity or the usefulness of such analyses. I cite them only to point out how the existence of emergent coding adds a new level of complexity to the analysis of human social behavior and of human culture, because behavioral interactions among people or groups of people can also change the codes that drive the behavioral interactions.

Any deeper analysis of complexity theory is beyond the scope of this book. Suffice it to say that it is a fundamental finding of complexity theory that feedback produces nonlinearity, change that is not additive but exponential. Nonlinear processes turn out in many cases to be impossible to predict, at least in detail. Since emergent coding involves two kinds of feedback – among individual agents and between behavior and coding – there is a double nonlinearity that merits analysis in terms of the mathematics of nonlinear systems.

2.4. CONCLUSION

In summary, then, human culture consists of the following:

1. Codes are created and maintained or modified through social interactions among individuals. These codes, along with genetically determined, individually learned, and socially learned (memetic) codes, govern the behavior of individuals. However, cultural codes are emergent, in that they cannot be understood without understanding the interactions that created them.
2. Among all living humans, socially constructed codes appear to motivate as well as inform the behavior of individuals. This means that individuals may be led to behave in ways detrimental to their own individual evolutionary fitness.
3. Cultural codes, in all present-day societies, have come to form ubiquitous, all-encompassing systems that assign cultural meaning, value, and so forth to almost everything that humans perceive, think, or do.

When I use the term "human culture," I do so because these three items describe what culture is like for all humans today. It is possible that we share certain aspects of culture with other primates or even nonprimate species. Certainly we share the propensity to learn from conspecifics and to create local traditions with other species, but this is not what I mean when I use the word "culture." In chapter 4, I review the currently available primate literature, looking for evidence that we share any of the three aspects of culture just listed. The results are, in my estimation, negative, but either new data or a more sophisticated reading of the existing data may change this. Moreover, it is possible that some, perhaps many, of our hominin ancestors lacked one or more of these three aspects of modern human culture. One thing I do later is to try to pinpoint in time the origins of each of these three phenomena.

Another thing I attempt to do, starting in the next chapter, is to formulate hypotheses to explain each of these three aspects of culture. I then evaluate these hypotheses in view of the available archaeological and fossil data.

3

WHY DOES CULTURE EXIST?

It may seem strange to ask why culture exists, given how central it is to human life and how well it has served our species. Yet the relationship between culture and human adaptation is more complex than it might at first appear. For one thing, unless we assume a priori that culture as we know it is a monolithic phenomenon with a single origin and a single explanation, we must in fact explain each of its components:

- Socially constructed coding
- Motivation of individual behavior by socially constructed coding
- Culture as ubiquitous and all encompassing

In this chapter, I try to propose plausible hypotheses to account for each of these. Testing them, however, is not a simple matter, because it involves collecting and evaluating data from disparate disciplines. For this reason, I as an archaeologist am able to test only the hypotheses concerning the elaboration of socially constructed coding into all-encompassing cultural systems.

3.1. EXPLAINING SOCIALLY CONSTRUCTED CODING

The first component of human culture, the social creation of coding, is by far the easiest to explain. In certain circumstances, all members of a group may find it in their individual self-interest to cooperate. For example, wolves often hunt animals such as moose (*Alces alces*) that are much larger than they are and that present a real danger to solitary hunters. They do so by attacking and mobbing the victim in packs. When hunting calves, wolves may perform complementary tasks: some will harass the mother while others await an opportunity to attack the calf. All share in the kill (Mech 1970:205-220). Clearly, this is a situation in which all the individual wolves benefit. They are able to kill an animal that, individually, they would have trouble killing, and the carcass is large enough that all members of the pack have something to eat.

It is easy to imagine that such a cooperative hunt would be more efficient if the wolves were better organized and had a more effective manner of communication. For a group of early humans, simple, socially constructed rules for how to carry out a cooperative hunt would benefit all participants. So, too, would a simple language that permitted better communication among the hunters – and language, of course, is one form of socially constructed coding. Thus, as long as the individuals involved in the hunt were motivated by their private internal coding, then the mere fact of cooperation raises no issues in evolutionary theory, even if that cooperation were enhanced by the use of socially constructed coding.[*]

It is possible that at some point in hominin evolution, socially constructed coding was used in this way, to inform – but not to motivate – people's behavior. Motivation came from the same kinds of codes – genetic, learned, or memetic – that motivate cooperation in other species. In other words, even when allowing emergent, socially constructed codes to inform and guide their behavior, individuals were acting in their own "perceived" evolutionary interests. (By "perceived" I do not mean conscious evaluation of the evolutionary implications. I simply mean that the socially constructed codes were in harmony with the genetically determined codes that produce adaptive behavior.) External coding that went against such perceived evolutionary interests would have been rejected by the individual, just as memetic coding that was perceived as maladaptive would have been rejected.

It is only when culture creates coding for *motivation* that the issue arises of possible conflict between the evolutionary interests of the individual and those of the group – when, for example, religion promises immortality and eternal bliss in return for evolutionarily maladaptive behavior such as chastity, or eternal suffering as a punishment for evolutionarily adaptive behavior such as giving in to lust.

If motivation is not affected, it is easy to find adaptive advantages of socially constructed coding. This is especially true because human language is a form of socially constructed coding. Any adaptive behavior that could benefit from either better communication or better coordination among individuals can serve as a potential explanation for the origins of emergent coding. This would include cooperative activities such as hunting. It would include teaching one's offspring. It would also include behaviors not found in other mammals. For example, a group might enhance its chances of finding food by dividing into several small

---

[*] I am not arguing that socially constructed coding originated in cooperative hunting. This is merely an example of the kinds of cooperative activities that would benefit from emergent coding.

foraging parties and agreeing to meet at a specific location and share either food or information.

Finding plausible hypotheses is easy, but two questions are more difficult to answer. The first is to determine which of all the plausible hypotheses is the actual, empirically, historically correct explanation for the origins of emergent coding. I do not attempt to provide a comprehensive list of possible hypotheses or to compare such a list with the empirical data in order to choose the best fit. The task would have been enormous. As early as 1975, a bibliography of writings on the origins of language (which is only one subset of socially constructed coding) ran to two volumes (Hewes 1975). It is more useful in this book to see if there is a way to pinpoint chronologically the origin of emergent coding.

The second question is why, given that its benefits seem obvious, socially constructed coding is not more widespread, at least among classes such as mammals or birds that have a highly developed ability to learn new behaviors. One possible answer might be that the creation of such coding requires a great deal of intelligence. If we measure intelligence, as Jerison (1982) suggests, by the allometrically adjusted ratio of brain size to body weight, then the cost of emergent coding would be, according to this explanation, the cost of maintaining a large brain (Aiello and Wheeler 1995; Foley and Lee 1991). This might indicate that something else, such as the development of a complex material culture or social competition (Alexander 1989; Byrne and Whiten 1988; Dunbar 1998), could have been the prime cause, and that the cognitive ability needed to create coding socially was merely a by-product of the evolution of some other adaptation.

The specific circumstances that produced socially constructed coding as an adaptation may still be in question. The general questions of how it could be adaptive and how it could evolve through natural selection, however, are easy to answer. For this reason, I pursue the matter no further, although in chapter 4 I attempt to date as closely as possible the origins of socially emergent coding. This requires reviewing the primatological evidence to determine as best I can whether or not this is an adaptation humans share with other primates. Since this appears not to be the case, I then turn to the archaeological and skeletal records to seek its origins in our own hominin lineage.

3.2. EXPLAINING WHY HUMANS PERMIT SOCIALLY CREATED CODING TO MOTIVATE THEIR BEHAVIOR

As soon as socially constructed coding provides a motivation for behavior, the possibility arises that the group (however defined) will create coding that will motivate individuals to behave contrary to their individual evolutionary interests. Thus we must find an explanation for how a willingness to let external codes motivate our behavior could have evolved. The problem is this. Individuals whose genotypes incline them to accept cultural coding that decreases their fitness will contribute fewer offspring to the next generation than will those whose genotypes incline them to selfishness. As a result, any such tendency should be destroyed by natural selection.

3.2.1. Altruism: The Individual versus the Group

This problem has been extensively investigated by those concerned with the origins of altruism – behavior that benefits the group at the expense of the individual. The topic has been a focus of research among evolutionary biologists, psychologists, sociobiologists, and some anthropologists. It is not my purpose in this book to solve or to review in detail any of the current theoretical debates concerning altruism.[*] However, I do need to introduce certain concepts from this theoretical literature that are relevant for explaining culture.

There are only two ways around the dilemma posed by altruism. Either altruism never evolved and altruistic behavior does not exist, or else it evolved because in certain circumstances natural selection favors altruistic behavior to an extent that overcomes negative selection at the individual level. I first discuss exactly how altruism must be defined in an evolutionary analysis. I then discuss how some evolutionary theorists believe that altruistic behavior could survive natural selection. Finally, I examine alternative hypotheses for our willingness to let culture motivate our behavior, hypotheses that do not invoke altruism in the evolutionary sense.

It is essential for our purposes that altruism be defined in evolutionary, not psychological, social, or philosophical terms. That is, it must be defined in terms of its consequences for the evolutionary fitness of the helper and the helped. This is a different set of criteria from those used in other contexts. For example, Batson (1991) wrote that

> even when we help in the absence of obvious external rewards, we may still benefit. We can receive self-rewards, congratulating ourselves for being kind and caring, or we can avoid self-censure, escaping shame and guilt. In such cases, the pat on the back may come from ourselves rather than from someone else, but it is a pat nonetheless.

---

[*] Sober and Wilson (1998) provide a good review of the literature, albeit from one particular point of view.

Such concerns may be of significance to the moralist or the psychologist, but from an evolutionary perspective what matters are the consequences a behavior has for the evolutionary fitness of the individuals involved – for their relative probabilities of successful reproduction.

From this perspective, behavior that helps others can be divided into three categories (Smuts 1999; Wilson 1990; following Wilson 1997):

1. Individual behavior that benefits the individual and all other members of the group (however defined) more or less equally
2. Individual behavior that benefits the individual but also benefits other members of the group more than the individual
3. Individual behavior that benefits the group at the expense of the individual

We can add to this list a fourth definition:

4. Individual behavior that is detrimental to the individual's fitness as an individual but that benefits the group, including the individual as a member of the group (Sober and Wilson 1998)

Clearly, behavior of the first kind is not difficult to explain in terms of individual fitness. It does not qualify as altruism in evolutionary terms because there is no sacrifice involved; it does not decrease the individual's fitness. Behavior of type 3 does decrease the individual's fitness and therefore must be considered altruistic in an evolutionary sense. Behavior of type 4 may decrease or increase the individual's fitness, depending on the relative effects on the fitness of the individual as a helper and as a member of the group. Behaviors of types 3 and 4 must be explained, either by some form of group selection or by the null hypothesis, that no such behavior actually exists. Behavior of type 2 is a borderline case. It increases the individual's absolute fitness, but it decreases his or her relative fitness.

3.2.1.1. Distinguishing between Real and Apparent Altruism.

There are certain kinds of behavior that on the surface appear altruistic but that actually enhance the fitness of the individual. In addition to cooperation, these have been described under the rubrics of nepotism and reciprocity.

3.2.1.1a. Cooperation. Deutsch (1949:133) defined cooperation as a situation in which all individuals involved benefit because their goals not only coincide but complement each other:

> If A, B, C, etc., does not obtain his goal..., X does not obtain his goal.

X obtains his goal only if A, B, C, etc., obtain theirs.
A, B, C, etc., obtain their goals only if X obtains his.

Such cooperation does not qualify as altruism in the evolutionary sense, because each individual receives an immediate net benefit. For example, in the case of cooperative hunting by wolves, every wolf is better off if it cooperates than if it does not. There is therefore no theoretical difficulty in explaining how a propensity toward this kind of behavior could evolve through natural selection.

3.2.1.1b. Nepotism and Inclusive Fitness. In general, mammals (especially females) expend a good deal of individual effort caring for their young offspring. Although such behavior requires sacrifice on the part of the parent, it cannot be considered altruistic in an evolutionary sense, because it is an adaptation by which mammals maximize the probability that their offspring will survive. In other words, they are maximizing their chances of successful reproduction.

Hamilton (1964) pointed out that in fact fitness should be measured not in terms of one's own individual reproduction per se but in terms of the likelihood that one's genes will be passed on to future generations. Since one shares a large portion of one's genotype with close relatives (50% with a sibling, 25% with a sibling's offspring, etc.), behavior that enhances their reproductive success will, ipso facto, enhance one's own "inclusive" fitness. Thus, behavior that (in probabilistic terms) may be costly for the individual is explainable as long as the cost is outweighed by the probable benefits for close relatives. As Haldane put it, "it would make sense for him to lay down his life to save the lives of eight of his cousins – since first cousins, on average, have one-eighth of their genetic material in common" (from Frank 1988:25).

Thus, seen from one theoretical perspective, the concept of inclusive fitness places one kind of self-sacrificing behavior outside the category of evolutionarily defined altruism. Such behavior should be termed nepotism rather than altruism. It should be noted, however, that other theorists (e.g., Sober and Wilson 1998) argue that in fact nepotism is simply one kind of group selection (see below).

3.2.1.1c. Reciprocity. Trivers (1971) pointed out that another form of apparently altruistic behavior can actually be explained in terms of an individual's self-interest. It can benefit an individual in the long term to make sacrifices for another individual (or individuals) as long as he or she has good reason to expect that in the future the other individual or individuals will reciprocate. Such behavior can be considered a form of delayed cooperation. However, it works only if there is some reason to expect reciprocity.

There is a large literature on the subject, much of it based on game theory. Essentially, all of these investigations begin with the prisoner's dilemma scenario (Axelrod 1984). Two prisoners (A and B) are being interrogated separately. Each must choose whether to give evidence against the other prisoner or to remain silent. Values are assigned to the consequences for A if he cooperates (remains mute) or defects (testifies against B), as shown in Table 3.1.

Table 3.1 Years imprisonment for A, depending on cooperation or defection of A and B.

	If B cooperates	If B defects
If A cooperates	1	5
If A defects	0	3

Clearly, A's best choice is to defect — but only if B cooperates. If both defect, then the results are unpleasant for both. However, if both cooperate, their jail time will be relatively short. (Here the payoffs are couched in negative terms, but they work the same way if stated in terms of positive benefits rather than punishments.) Computer simulations showed that in the game of prisoner's dilemma, the best strategy – the one that produced the highest rewards for A over many repeated interactions with B – was one called tit-for-tat, in which player A first cooperates. If player B defects, then player A defects and continues to do so until player B cooperates, at which point A again cooperates and continues to cooperate as long as B does so as well.

This scenario, its analysis in terms of game theory, and computer modeling and simulation have been refined and modified since Axelrod's original publication (Axelrod 1997; Axelrod and Dion 1988; Cox et al. 1999; Trivers 1985), but the implications remain essentially unchanged. Reciprocity of this kind can be explained in terms of benefits to an individual's fitness, without recourse to group selection. It can also be explained in terms of individual coding, without recourse to socially constructed coding – although of course it cannot be explained without recourse to social interaction.

Alexander (1987) argued that another kind of reciprocity was possible beyond direct exchange of help between two individuals. Returns to one who provides help may come not from the individual helped but from others. Alexander's definition of indirect reciprocity is fairly broad, but for our purposes, the gist of it is explained in his Table 2.2:

A helps B
B helps (or overhelps) A

C, observing, helps B, expecting that
B will also help (or overhelp) C

Thus indirect reciprocity differs from direct reciprocity in that it involves the "reputation" one acquires among observers as a result of one's behavior. Nevertheless, the essential reason for helping others remains the same, the expectation that a helpful act will be repaid in the future. This expectation need not depend on any kind of emergent social coding.

3.2.2. Group and Multilevel Selection Explanations

To some sociobiologists and evolutionary biologists, all apparently altruistic behavior can be explained in terms of one or another of the foregoing categories. Others disagree, and argue that some form of group selection makes altruistic behavior not only possible but plausible.

The subjects of group selection and of the evolution of altruism have aroused a great deal of debate (e.g., Aoki 1982; Batson 1991; Boehm 1999b; 1999a; Boorman and Levitt 1980; Boyd et al. 2003; Boyd and Richerson 1982; papers in Brandon and Burian 1984; de Waal 1996; Dugatkin 1999; Falk 1985; Goodenough 1995; Hull 1981; Keller 1998; Maynard Smith 1964; 1976; Pepper and Smuts 2000; Reeve and Keller 1999; Smuts 1999; Sober and Wilson 1998; Soltis et al. 1995; Wilson 2002; Wilson and Kniffen 1999; E. O. Wilson 1975; Wonderly 1996; Wuketits 1996; Wynne-Edwards 1962; 1986). The notion that natural selection can operate on groups as well as on individuals has gone through cycles of popularity and unpopularity (see Reeve and Keller 1999; also see Sober and Wilson 1998 for a review). Without some form of group selection, true altruism could not evolve. Instead, all apparently altruistic behavior would have to be explained in other terms, including nepotism and reciprocity.

There are really two versions of group selection theory. The first explains the evolution of individual behavior that decreases the fitness of the individual but increases the fitness of other members of the group. Such behavior should be eliminated by natural selection operating *against* individuals with a genetic propensity for altruistic behavior. There is a considerable literature analyzing the conditions under which group selection can, theoretically, overcome this effect (e.g., Aoki 1982; Boorman and Levitt 1980; Keller 1998; Maynard Smith 1964; 1976; Pepper and Smuts 2000; E. O. Wilson 1975). Generally speaking, it seems to be agreed that:

- Groups must be internally rather homogeneous – altruistic individuals must tend to group together. (Living in proximity to relatives is one mechanism by which this can be achieved, which is why some theorists consider kin selection to be a form of group selection.)

- Between-group diversity must be high.
- Gene flow between groups must be low.
- Extinction rates for groups must be high.

Two additional considerations should be noted. First, competition among groups need not involve outright extinction of entire groups. Rather, competition may involve differential reproduction, meaning differential contributions to the next generation (Maynard Smith 1964; Sober and Wilson 1998; Williams and Williams 1957).

Second, a number of theorists have invoked moral or cultural norms as factors that increase within-group behavioral homogeneity (e.g., Axelrod 1986; Boehm 1999b; 1999a; Sober and Wilson 1998:149-156; see also Boyd and Richerson 1982 and Wilson and Kniffen 1999 for similar arguments in a memetic context). The nature of these norms is variously defined and explained. For our purposes, however, using cultural norms to explain group selection would create a circular argument – we would be using motivation by cultural norms to explain why humans permit cultural norms to motivate their behavior.

The requirements of this first version of group selection are rigorous. A second version, usually called "multilevel" selection, has been developed more recently. It sees altruism as individual behavior that reduces the fitness of the individual relative to other members of the group but increases the fitness of all members of the group (including the altruistic individual) relative to members of other groups.

The multilevel selection model, developed primarily by Wilson (Sober and Wilson 1998; D. S. Wilson 1975; 1997), takes its name from a general view that selection operates on several levels, including the gene, the individual organism, the group, and the species. From this perspective, the individual has no special status. In terms of individual versus group selection, which is what concerns us here, within-group selection acts against individuals who act altruistically. At the same time, between-group selection favors groups containing altruists. The altruistic individual benefits from the latter, along with everyone else. The sum total of selective forces must be measured in terms of both within-group and between-group components. (This is analogous to a balanced polymorphism, where a given allele has both deleterious and beneficial effects on fitness.) The actual results depend on the strength of each, as well as the pattern of within-group and between-group phenotypic variation and the degree of heritability of phenotypic variations (Sober and Wilson 1998:101-116).

If we consider the willingness or susceptibility to be motivated by cultural codes as a form of altruism – or as a proximate mechanism for altruistic behavior – then either group or multilevel selection can be proposed as a hypothesis to explain its evolution. There is still no consensus, however, within evolutionary biology and evolutionary theory about the plausibility of group or multilevel selection. How the debate is eventually resolved will have real consequences for the way human culture is explained.

Nevertheless, other hypotheses are available that may account for our willingness to let culture motivate our behavior. In each of these, such a tendency would have evolved because it enhanced the fitness of the individual. Any negative effects on fitness would have been more than offset by positive ones.

3.2.3. Frank's Commitment Hypothesis

Economic theory relies heavily on the assumption of rational behavior, yet it is apparent that emotions lead people to behave in ways contrary to this assumption. In response, one economist has developed an explanation similar to Alexander's notion of indirect reciprocity (Frank 1988). Vengeful behavior, for example, cannot be explained in terms of rational cost-benefit analysis. Frank uses the example of the famous nineteenth-century blood feud between the Hatfield and McCoy families in Kentucky and West Virginia. Over a period of 35 years, numerous members of both families were killed. Frank points out that in rational terms it would have been beneficial to either side to have broken the cycle of violence, yet neither did so. Humans seem to have an emotional propensity to react to hostile behavior in ways that seemingly cannot be accounted for in terms of calculated self-interest.

Frank explains this as a mechanism for preventing exploitation by others:

> Suppose Smith grows wheat and Jones raises cattle on adjacent plots of land. Jones is liable for whatever damage his steers do to Smith's wheat. He can prevent damage altogether by fencing his land, which would cost him $200. If he leaves his land unfenced, his steers will eat $1000 worth of wheat. Jones knows, however, that if his steers do eat Smith's wheat, it will cost Smith $2000 to take him to court. (Frank 1988:48)

If Smith were economically rational, he would not sue, and it would then not be rational for Jones to fence his land. However, if Jones knows that Smith's emotional makeup is such that he will sue regardless of cost, then Jones will fence his land, thereby saving Smith $1,000 worth of wheat.

Thus, it may be useful to the individual to have a reputation for emotions that would otherwise seem irrational. Such emotions are a commitment to behavior that cannot be explained in terms of short-term cost-benefit analysis. There are two grounds for such a reputation. On the one hand, being observed behaving vengefully will give one a reputation for vengefulness. On the other hand, an emotional tendency toward vengefulness may be visible to others even in the absence of vengeful behavior. Franks cites empirical data that tend to support this.

Vengefulness is a negative example of indirect reciprocity, but Franks points out that a reputation for positive traits such as generosity can also be useful, by inducing trust and reciprocity from others. Overall, then, indirect reciprocity and Franks's commitment theory are similar, at least in the context of this book. Both explain apparently maladaptive behavior, including apparent altruism, in terms of benefits to the individual. It could also explain why people might be motivated, genetically, to adhere to socially constructed coding for behavior that appears irrational or evolutionarily disadvantageous in the short term. A reputation for being willing to abide by socially constructed norms and values could bring the individual sufficient benefits over the long term to offset his or her sacrifices.

3.2.4. Simon's Docility Hypothesis

Simon's docility hypothesis was formulated in the context of socially learned traditions. Because the "meme pool" of any human group is liable to be dominated by memes that are appropriate for the group's environment, migrants into the group are liable to bring with them memes that are unsuitable to the new environment. Therefore, a "conformist transmission rule of the form 'when in Rome do as the Romans do' is advantageous in this circumstance" (Richerson and Boyd 1998). Simon (1990) called this tendency "docility." He pointed out that although docility is advantageous because individuals reap the benefits of others' experience and knowledge, the benefits will be partially offset if docility also leads them to behave altruistically. Overall, though, natural selection (at the individual level) can still favor docile individuals who behave altruistically because of their docility. I would add that this is similar to the evolutionary consequences of learning. Overall, learning is beneficial. On occasion, our ability to rewire our brains may lead to maladaptive behavior, but in statistical terms this is more than compensated for by the flexibility it provides.

Simon's and Richerson and Boyd's hypothesis can serve to explain how individuals would permit socially constructed coding to motivate their behavior, even to the point of altruism.[*]

---

[*] Simon goes on to argue that altruism is beneficial at the level of between-group selection. Thus his argument can also be seen as a form of multilevel selection model, with

3.2.5. A Hypothesis Invoking Individuals' Adaptation to Culture as Environment

Flinn and Alexander (1982:397) wrote that "culture can be regarded as an aspect of the environment into which each human is born and must succeed or fail, developed gradually by the succession of humans who have lived throughout history." Their concept of culture was essentially memetic rather than emergent. Nevertheless, there is no question that culture, even as I define it, is a part of every individual's environment, and that success in human life requires adjusting to that culture. It can therefore be hypothesized that our willingness to let cultural coding motivate our behavior is simply an individually adaptive response to the existence of socially constructed coding.

This hypothesis is plausible if systems of socially constructed coding evolved that improved the effectiveness and efficiency of cooperative behavior. Such systems could come to be of considerable importance to a group's adaptation without requiring altruism on the part of any of the group's members.

If such systems came to rely to any significant degree on reciprocity rather than pure cooperation, then they would have to solve the problem of cheating. One way to accomplish this is simply not to interact with cheaters. The problem with this solution is that the benefits of reciprocity are lost. Another way is to punish them (Axelrod 1986; Boyd and Richerson 1992; Maynard Smith 1999), but if the act of punishing has any cost for the punisher (e.g., a danger of injury or retaliation), then the act of punishing is itself a form of altruism (Sober and Wilson 1998:143-144).

However, if a system based on socially constructed coding – for example, procedures for communal hunting or arrangements for collecting and sharing information about resources – becomes important enough to an entire group, and if it relies heavily on cooperation by all concerned, then everyone in the group will have an incentive to punish cheaters that may outweigh the costs of doing so. In other words, the benefits, even to the individual punisher, of punishing cheaters may outweigh the costs. (This would be especially true in small groups such as those presumably common in the Paleolithic.)

Under such circumstances, the costs of cheating would quickly outweigh the costs of cooperating, and a tendency to adhere to the demands of socially constructed coding could be adaptive. In this case, the system of cultural coding, along with the willingness of others to punish cheat-

---

the benefits of docility, to the individual, added to the balance of within-group versus between-group selection.

ers, would be an aspect of the environment to which members of the population would have to adapt..

3.2.6. Testing the Hypotheses

The origins of our willingness to let culture motivate our behavior lies somewhere in the past. It would seem logical, therefore, to turn to the historical sciences – archaeology and paleontology – for evidence to test these alternative hypotheses. However, I can think of no archaeological or skeletal correlates of such motivation. If this is correct, then cultural motivation would be archaeologically and paleontologically invisible.

It is true that the social construction of coding may have archaeological and paleontological correlates. For example, changes in brain morphology related to the origins of language may be reflected in cranial endocasts. (See chapter 4 for a discussion of this and other possible correlates.) However, unless one assumes a priori that creation of and motivation by cultural codes appeared simultaneously, then such data still tell us nothing about the origins of the latter phenomenon.

If this is so, then the historical sciences will be unable to test the relative merits of the foregoing hypotheses. This task must fall to other disciplines that deal with living subjects. As an archaeologist, I will not attempt it.

Yet the matter is still of relevance to paleoanthropologists. The elaboration of socially created coding into all-encompassing cultural systems is a development that, I believe, is quite visible archaeologically. One hypothesis to explain this development requires that when this elaboration occurred hominins were susceptible to cultural motivations.

3.3. EXPLAINING THE ELABORATION OF CULTURE

We must explain why socially constructed coding developed beyond the point necessary to increase the efficiency and effectiveness of teaching offspring, of cooperative activities, and the like. Why has culture developed into a ubiquitous and all-encompassing framework that incorporates virtually everything we perceive, think, or do, including things that have little or nothing to do with cooperative endeavors? Why do we construct myriad things (supernatural beings, concepts, values, cultural categories, etc.) that are entirely cultural and have nothing to do (at least directly), with the practical aspects of cooperative activities? (For the sake of succinctness, I use the shorthand term "elaboration of culture" to refer to this phenomenon.)

In the remainder of this chapter, I propose a series of alternative hypotheses to account for these phenomena. I also suggest, when possible, archaeological test implications for each of the hypotheses. In chapter 5, I review the archaeological evidence in light of these test implications. In general terms, the hypotheses fall into three categories:

- The elaboration of culture may be explained as a by-product or "spandrel" (Gould and Lewontin 1979) of the evolution of the mental abilities required for socially constructed coding. The most likely of these would be the neural structures that make language possible (language being a fundamental form of socially constructed coding). The elaboration of culture would be adaptively neutral.
- The elaboration of culture may be explained in terms of its evolutionary benefits for the individual.
- The elaboration of culture may be explained in terms of its evolutionary benefits for the group (however defined), even at the expense of individual fitness.

3.3.1. Explaining Cultural Elaboration as Adaptively Neutral

The "By-product Hypothesis": The elaboration of culture is an adaptively neutral spandrel or by-product of the evolution of socially constructed coding.

It will be seen in the next chapter that socially constructed coding plays either no role or at most an extremely limited role in the adaptations of other primates. This is probably because crucial cognitive abilities are insufficiently developed in those species, and the most important of these is probably related to language. Language facilitates the social construction of other coding and permits the construction of much more elaborate coding than would be possible without it. The great apes seem to differ from humans only quantitatively in the development of key linguistic abilities, especially in the ability to understand and use symbols. Nevertheless, these quantitative differences are real, and there is no good evidence that chimpanzees in the wild make use of the kind of arbitrary symbolic reference that underlies human language. As will also be seen in chapter 4, the reorganization of the brain toward its modern human state began as far back as *Australopithecus* and accelerated throughout the evolution of *Homo*. It is possible that a quantitative increase in the mental abilities underlying the use of symbolism led to language and to effective social construction of other kinds of coding.

Symbolic ability would originally have evolved either in order to fulfill practical requirements for efficient communication or as a by-product of other mental abilities. However, once the hominin brain became adept at the use of symbols in communication, it is possible that symbols were used more or less indiscriminately, even in contexts where they were of little evolutionary benefit. The result could have been the elaboration of culture – a proliferation of culturally created symbolic entities and the incorporation of much of what hominids perceived, thought, or did into a symbolic matrix.

What are the archaeological test implications of such a hypothesis? If culture as an all-encompassing intellectual framework is nothing more than a by-product of the development of the symbolic abilities underlying human language, then evidence for language and evidence for culture as all-encompassing should appear essentially simultaneously, because the former would be the sole cause of the latter. This is probably the most useful test implication for evaluating this hypothesis archaeologically. In chapter 4 I review the data regarding the origins of language, and in chapter 5 I review the kinds of evidence that permit us to date the origins of culture as all-encompassing.

3.3.2. Explaining Cultural Elaboration by Adaptive Benefits to the Individual

An "Anxiety Hypothesis": The elaboration of culture can be explained as a mechanism for allaying individual emotional anxieties.

I have not found it easy to identify a compelling reason why individuals as individuals (rather than as members of a society) need cultural coding beyond the level of socially constructed coding sufficient to educate the young, coordinate cooperative group activities, and so forth. Yet there is no denying that individuals do use cultural ideas to allay their anxieties. An archaeologist – unfortunately I cannot remember who it was – once said to me that religion was necessary as soon as our ancestors understood that death was inevitable. There is certainly no question that religion eases our fear of death (except when concepts such as hell make it even more terrifying), as well as all our other fears, sorrows, and similar emotional burdens.

The history of functionalism in culture theory follows two strands, the group-oriented functionalism of scholars such as Radcliffe-Brown (1935/1952) and the individual-oriented functionalism of Malinowski, who formally proposed that culture (in the form of magic) served to allay anxiety (Malinowski 1944; 1954). Certainly there can be little doubt that culture, or at least certain cultural concepts, rituals, and artifacts, do function to allay anxieties and to provide emotional support to individuals.

It is possible to state this as a hypothesis for explaining the elaboration of culture: once language and symbolism were available to hominids, they used them to create a symbolic, cultural structure that served to allay individuals' anxieties and other emotional stresses. Although this may not have been a crucial function, as long as the costs of cultural elaboration were low, it could be accounted for by the benefits of reduction of psychological stress. Thus this hypothesis accounts for the expansion of cultural coding into an all-encompassing symbolic framework.

Since death existed long before the first primates appeared, and since anxiety and fear are hardly unique to humans, it follows logically that, as soon as the means for using culture to allay anxiety were available, this would have been done. If those means were language and symbolism, then we would expect that language and the elaboration of culture would appear essentially simultaneously.[*]

3.3.3. Explaining Cultural Elaboration by Adaptive Benefits for the Group

A "Group Benefit Hypothesis": The elaboration of culture is a means of motivating altruistic behavior that benefits the group, even at the expense of the individual.

This hypothesis explains the elaboration of human culture as a tool that society uses to overcome the natural reluctance of individuals to behave altruistically. More precisely, the elaboration of culture is a tool that society uses to enhance and reinforce an individual's willingness to let cultural coding motivate his or her behavior.

Recall that every individual's behavior is guided by several layers of coding, of which cultural coding is only one. Cultural codes must compete with other, individual-centered codes to guide a person's behavior. Cultural elaboration opposes the resulting tendency toward selfish behavior by providing an overarching intellectual and emotional framework within which individuals operate. This cultural system can motivate group-beneficial behavior in four ways (Chase 1999):

First, culture does not place demands on the individual without reinforcing those demands psychologically. It includes a set of rules, definitions, values, and so forth that define how the individual is expected to behave. It also involves a set of cultural concepts, usually embedded in mythology, that explain why this is behavior expected. Cultural imperatives, especially the most important ones, are not presented as meaningless, arbitrary rules. Rather, they are embedded in a worldview that legitimates them by contextualizing and explaining them.

One example of the way an elaborated cultural worldview both defines and justifies what the individual must do is provided by the

---

[*] Atran and Norenzayan (2004) made a very similar argument, although they postulated a somewhat different means, namely, the capacity for impossible beliefs.

Murngin of Australia, who were described by Warner (1937/1958) when they were living by hunting and gathering in an environment marked by extreme seasonal fluctuations in rainfall. These seasonal changes were stressful but essential to the Murngin adaptation. The Murngin world-view incorporated the seasons, the entire natural environment, and the social structure in a system of principles, of institutions such as the moiety and clan, of obligations, and of prohibitions expressed in mythology and supported by ritual. Indeed, ritual was believed necessary to the proper operation of the natural as well as the social world.

> That which organizes the two categories (the seasonal reproductive cycle of nature and the male-cleanliness female-impurity dichotomy of society) into one is the totemic symbol which can be manipulated by ritual and gives man an effective control over nature and an effective negative sanction over members of his society. The rituals must be properly conducted yearly to keep the group and its individuals ritually clean; and in these rituals the manipulation of the sacred totem insures the proper function of the seasons, a sufficient production of food, and a continuation of the natural surroundings proper for man. Thus that which is beyond man's technology or beyond his real powers of control becomes capable of manipulation because its symbols can be controlled and manipulated by the extraordinary powers of man's rituals. At the same time the identification, in the totemic concept, of the male and female principles with the seasonal cycles gives the adult men's group the necessary power to enforce its sanctions; the providing world of nature will not function if the rules of society are flouted and man's uncleanliness contaminates nature. Hence everyone must obey. If he does not by his own volition, then he must be forced to. (Warner 1937/1958:396)

The second way in which the cultural system can motivate group-beneficial behavior, in addition to justifying and explaining its demands, is by providing society and the individuals in society with a reason to enforce its dictates, by punishment if necessary. In species without the kind of culture we find among humans, individuals are not motivated to interfere with the behavior of other individuals, except when it threatens their self-interest, either directly or indirectly through a relative or ally. In the context of elaborated cultural systems, however, as we just saw in the Murngin example, not only are there rules that must be obeyed for purely cultural reasons, but there are rules that must be enforced as well. A Murngin need not injure another person's welfare in any concrete manner in order to be punished; any violation of cultural norms is a threat to the seasonal cycle and therefore to society as a whole, and must be sanctioned.

Third, in addition to motivating members of society to punish those who violate cultural norms, culture may provide its own rewards and punishments more directly, at least in the minds of those who accept its

worldview. For example, Stefánsson (1913/1962:272) described an Eskimo hunter who, having failed to notify his fellow hunters when he
killed a bearded seal and having kept all the meat for himself, felt bitter
remorse and blamed on his selfishness the blindness that later befell him.
Thus cultural concepts such as salvation and damnation provide strong
motivations for the individual to adhere to cultural norms.

Finally, culture uses the medium of ritual to reinforce its values emo-
tionally. Gellner (1988:130) put this very colorfully:

> The way in which you restrain people from doing a wide variety of
> things, not compatible with the social order of which they are mem-
> bers, is that you subject them to ritual. The process is simple: you
> make them dance round a totem pole, until they are wild with ex-
> citement, and become jellies in the hysteria of collective frenzy, you
> enhance their emotional state by any device, by all locally available
> audio-visual aids, drugs, dance, music, and so on; and once they are
> really high, you stamp upon their minds the type of concept or notion
> to which they subsequently become enslaved. The ... central role of
> ritual is the endowment of individuals with compulsive concepts
> which simultaneously define their social and natural world and con-
> trol their perceptions and comportment in mutually reinforcing ways.

His description, though, is overdramatic. Less colorful rituals, such as
saying grace before meals or pledging allegiance to the nation's flag
every day, are much more common and probably have a greater overall
effect.

There are two kinds of tests that can be used to evaluate the validity
of the group benefit hypothesis:

1. The hypothesis requires that people today do behave altruistically, and that they do so – at least in part – not out of a general-
   ized urge to altruism but out of a propensity to obey the dictates
   of culture. Whether this is so cannot be tested archaeologically.
   It will have to be tested in other disciplines that deal with living
   people whose behavior can be observed directly.
2. In terms of the archaeological record, the group benefit hypothesis is consistent with the passage of considerable time between
   the first evidence of socially constructed coding and the first evidence of cultural elaboration. Such a time lag is inconsistent with
   either of the two alternative hypotheses discussed earlier.

From the perspective of the group benefit hypothesis, neither the
cognitive capacity for nor the existence of socially constructed coding
automatically produces cultural elaboration. Both are necessary but not
sufficient conditions. Some other specific conditions must provoke further change. I have argued that this could have happened, for example, in

circumstances where exploiting clumped, mobile, and unpredictable resources required cooperation over large areas of low population density, where opportunities for cheating would nullify the normal advantages of reciprocity (Chase 1999). The elaboration of culture would have occurred when (1) the ability to create coding through social interaction had evolved, *and* when (2) circumstances were such that it became beneficial for a group or groups of people to use such coding to motivate their individual members to behave altruistically, *and* when (3) individuals were for some reason willing to accept such motivation.

3.3.3.a. An Implication of the Group Benefit Hypothesis. If, as this hypothesis suggests, culture serves to motivate and to shape altruism, then it implies that genetic evolution has failed to do so, either wholly or in part. There are two alternative possibilities regarding altruism:

1. No tendency toward altruistic behavior per se evolved through genetic group or multilevel selection. Only culture, and especially elaborated culture, motivates whatever altruistic behavior occurs among humans. In other words, humans are characterized by a tendency to accept the dictates of culture, not by a tendency to behave altruistically.
2. Some tendency toward altruistic behavior evolved through genetic group or multilevel selection. Culture, especially elaborated culture, also motivates altruistic behavior. Altruistic behavior observed among humans can be explained by a combination of the two motivations.

3.3.4. Summary of the Archaeological Test Implications

If I am right in my belief that culture could motivate individual behavior without producing visible archaeological correlates, then archaeologists are in no position to evaluate hypotheses pertaining to the origins of this phenomenon. Such testing will have to be left to disciplines that draw their empirical data from living humans. Archaeology is, however, in a position to evaluate the relative merits of hypotheses generated to explain the elaboration of culture into overarching and all-encompassing systems of thought. I have proposed three such hypotheses:

1. The by-product hypothesis explains cultural elaboration as a by-product of the cognitive and symbolic abilities underlying socially constructed coding in general and language in particular. If it is true, then the elaboration of culture should have occurred as soon as these abilities evolved. If a significant time lag is evi-

dent, then the elaboration of culture must be explained otherwise.

2. The anxiety hypothesis explains cultural elaboration as a mechanism for allaying individual anxieties. If it is true, then the elaboration of culture should have occurred as soon as the cognitive and symbolic abilities underlying socially constructed coding and language were present. This is because anxiety and fear are constants. They are present in other primate species. Thus, if a significant time lag is apparent, then another explanation for the elaboration of culture must be invoked.

3. The group benefit hypothesis is compatible with a considerable time gap between the origins of socially constructed coding, including language, on the one hand, and evidence for the elaboration of culture, on the other.

In short, we can choose between the group benefit hypothesis and the other two hypotheses by examining the primatological, archaeological, and paleontological records to determine whether or not there was a time lag between the origins of socially created coding and the elaboration of such coding into all-encompassing systems.

In chapter 4, I review the evidence for the origins of socially constructed coding. The most visible, or at least potentially visible, form of such coding is language. Simple referential language, stripped of all its additional cultural (e.g., ritual, mythological, legal) aspects, is a fundamental form of socially created coding. The origins of language have been assiduously researched for well over a century. Nevertheless, it will be seen that the archaeological and paleontological data bearing on its date of origin are equivocal, largely because many of the bridging arguments involved remain insufficiently tested.

In chapter 5, I examine the archaeological record for evidence of the elaboration of culture. A number of phenomena in the archaeological record can indicate quite clearly when culture became – or was well on the way to becoming – an all-encompassing phenomenon that went beyond the practical needs of cooperation or communication:

1. *Evidence of mythology.* Through mythology, culture provides a worldview for all members of society. When we find depictions of mythological beings or other evidence of mythology, we can infer that culture had extended beyond the most basic level of socially constructed coding.

2. *Evidence of ritual.* By ritual I do not mean the stereotyped behavior found in other species, such as the courting behavior of peacocks or the greeting behavior of wolves. I mean instead behavior through which cultural ideas are acted out and cultural be-

liefs are reinforced. Burying the dead and cannibalism are often interpreted as evidence of this kind of behavior (but see chapter 5).

3. *Symbolic artifacts,* that is, artifacts whose only function, or one of whose major functions, appears to be symbolic or cultural in nature.
4. *Culturally determined style.* Evidence that artifacts were made to culturally defined standards beyond the practical considerations of manufacturing or physical function. This is generally referred to as style, but it must be clearly distinguished from the kind of regularity that is a natural product of noncultural, memetic traditions. As will be seen, this last requirement is problematic.

I will have a great deal more to say about these phenomena in chapter 5. How they are to be recognized in the archaeological record and how they are to be distinguished from similar but noncultural phenomena are complex matters. Laying aside practical considerations, however, we can accept them in theoretical terms as archaeological correlates of the elaboration of culture – indicators that culture had expanded to provide a framework of meanings and values far beyond the immediate requirements of coordinating practical activities or instructing children in practical tasks.

One more complicating factor needs to be considered. It is possible that cultural elaboration is not an all-or-nothing phenomenon, but that it developed gradually or even in sporadic, discontinuous bursts through space and time. Stringer and Gamble (1993:203) wrote: "We disagree ... that symbolic behaviour is something that can be turned up or down like a light on a dimmer switch. On the contrary, arranging behaviour according to symbolic codes is an all or nothing situation. The onset of symbolic behaviour can be compared to the flick of a switch." At the time, I agreed with this assessment, although I pointed out that the statement was unsupported either theoretically or empirically (Chase 1994). From my current theoretical perspective, arranging behavior according to symbolic codes really consists of three different phenomena that might or might not have evolved simultaneously. In addition, each of the hypotheses listed earlier has different implications for Stringer and Gamble's assertion.

The by-product hypothesis states that the elaboration of culture is the automatic and inevitable result of the brain's evolution beyond a certain threshold of symbol-using ability. The implication is that once this threshold was reached, the florescence of culture would have been almost instantaneous, at least when measured on a geological time scale.

The group benefit hypothesis depends on two things: a susceptibility to cultural motivation and a particular set of local circumstances under which it would be advantageous to a given group of people to take advantage of that susceptibility.

With regard to the first, as we have seen, there are a number of ways in which that susceptibility might have evolved independently of the actual use of that susceptibility for the benefit of the group. As for the second, I see no reason to reject a priori two possibilities:

1. Cultural elaboration as a way to motivate altruistic behavior might have arisen independently at different times, in different places, in response to local conditions. Once local circumstances changed, the use of cultural elaboration to motivate altruism might sometimes have disappeared.
2. In order to motivate altruism, culture would have to have been elaborated beyond the basic fact of socially created coding, but it would not necessarily have to have been elaborated to the extent that it is among humans today.

If these two scenarios held true, then the result would be a pattern of sporadic appearances and disappearances of archaeological traces of culture. Thus the group benefit hypothesis would be compatible with an apparently gradual* increase in the degree of cultural elaboration until it reached the stage observable among all humans today, the stage at which cultural concepts, values, and meanings incorporate almost everything we think or do.

The use of culture to allay anxiety is not incompatible with the use of culture to motivate altruistic behavior. The two can coexist. What I mean by the anxiety hypothesis, however, is that the phenomenon of cultural elaboration, as fully developed as it is among living humans, can be explained by its use to allay anxieties and other emotional stresses. The hypothesis posits only two prerequisites for fully elaborated culture: the necessary cognitive and symbolic capacities and the presence of anxieties and other emotional stresses. Since the second is present among all primates, fully elaborated culture should appear very shortly after the necessary cognitive and symbolic capacities.

Even if the anxiety hypothesis does not account for fully elaborated culture, it is possible that on occasion cultural codes (e.g., socially created belief in an afterlife) were used to allay anxieties even before culture was fully elaborated. In this case, however, a further cause would still have to be invoked to explain the full elaboration of culture; the anxiety hypothesis would be insufficient.

---

* Of course, the increase would not really be gradual in the sense of smooth or regular, but it would likely appear so given the coarse resolution of the archaeological record.

With these test implications in mind, we can turn to the evidence for the origins of simple socially constructed coding, on the one hand, and of the elaboration of culture, on the other.

4

THE ORIGINS OF SOCIALLY CONSTRUCTED
CODING

As I stated in the previous chapter, I will not investigate hypotheses formulated to explain the origins of socially constructed coding. However, there are at least three reasons to try to determine, to the extent possible, the date at which socially constructed coding originated. First, if it predates the separation between our lineage and that of other primates, then we share at least one aspect of human culture (as I am using the term) with other species. Second, because socially constructed coding is an important aspect of the human adaptation, any account of human evolution will be incomplete unless we know when it first appeared. Finally, the relative timing of the origins of socially constructed coding and of the elaboration of culture into an all-encompassing system is crucial to choosing among the alternative hypotheses proposed in the previous chapter.

In this chapter, therefore, I first review the primate data from both laboratory and field for evidence that other species create coding socially. I then look at the hominin paleontological and archaeological records.

Much of the evidence discussed in this chapter relates to the origins of language. There are two reasons for this. First, language is a form of socially constructed coding. It is important in its own right, and without it any but the very simplest forms of socially constructed coding would be impossible. Second, it may be the only kind of *unelaborated* socially constructed coding that may be visible archaeologically or skeletally. In part, this is because a huge amount of theoretical and empirical research has been done on the origins of language, but virtually no research has been directed toward the origins of socially constructed coding per se.

4.1. THE PRIMATE EVIDENCE

In examining the primate evidence, we cannot treat socially constructed coding as a monolithic phenomenon. Rather, we must break it down into various constituent parts:

- The cognitive ability to learn codes (rather than just behaviors) from conspecifics
- The cognitive and social ability to create codes through social interaction
- The actual use of socially created codes in the wild as part of the adaptive repertoire of a nonhuman species

If we possess all of these in common with other primates, then the use of socially constructed codes is part of the evolutionary history of the Hominidae (humans and great apes) or of a higher-level taxon. If we share none of these with the primates, then we must look to the fossil or archaeological record of hominin evolution for their origins. If we share some of them, then we must still look within hominid evolution for the origin of the remaining traits.

I must emphasize that the following pages are not intended as an overall review of the primatological literature or an overview of the way of life of the great apes. Their social life is remarkably rich and complex, and apes are remarkably intelligent, but it is not my purpose to document these facts. Rather, I comb the literature for data that might demonstrate – or even hint – that primates other than humans create coding through social interaction or are capable of doing so. What follows, therefore, is a very limited and specialized look at the behavior and cognition of nonhuman primates.

4.1.1. Ape Language Experiments

There is little in the ethological literature on memetic learning that indicates that primates or other mammals are learning *codes* from their conspecifics. As far as I am aware, almost nothing has been reported that cannot be explained either in terms of imitating another individual's behavior or in terms of observing another individual's behavior and in the process learning something about the environment that leads one to behave similarly. In other words, among nonhuman species, it appears to be behavior, not coding, that is "transmitted" from one individual to another. (Later in this chapter I discuss an exception, possible evidence of deliberate teaching among chimpanzees.)

In the laboratory, however, primatologists and psychologists have been making a concerted effort to teach apes to use at least a rudimentary form of human or humanlike language (Fouts and Budd 1979; Gardner and Gardner 1978; Gardner and Gardner 1969; Greenfield and Savage-Rumbaugh 1991; Hayes and Hayes 1951; Kellogg and Kellogg 1933; Miles 1990; Patterson 1978; Premack and Premack 1983; Ristau and Robbins 1982; Rumbaugh 1977; Savage-Rumbaugh 1986; 1987; Savage-Rumbaugh et al. 1986; Sevick and Savage-Rumbaugh 1994; Snowdon 1990). Since language is coding, this is a good place to look for evidence of a cognitive ability to learn codes from other individuals – albeit individuals of another species.

The languages used in these experiments include normal human speech (which apes are unable to produce because of the anatomy of their vocal tracts, but which they seem to understand), a variant of American Sign Language, and artificial languages based on plastic tokens or computerized "lexigrams." The simplest form of coding that the human experimenters have attempted to teach is the semantic symbol. By symbol I mean a sign (sound, gesture, token, or lexigram) that stands for or refers to something else by arbitrary convention (Peirce 1932 [1960]) – or, from the captive ape's point of view, by arbitrary human fiat. If the ape can grasp this aspect of a symbol, then it has learned a code. If, on the other hand, the ape simply learns, through a stimulus-response process, that there is a connection between the sign and a reward or punishment, then it has learned a behavior, not a code.

Strong doubts have been raised about whether apes have really learned symbols rather than behaviors, especially with regard to earlier experiments (Seidenberg and Petitto 1987; Terrace 1979; Terrace et al. 1979; Wallman 1992). Fortunately, questions raised by the skeptics have stimulated more rigorous investigations, and at present it seems that even rather skeptical experts accept that some apes have grasped the concept of a symbol (Snowdon 1990; Tomasello and Call 1997).

If this is in fact the case, then we share with at least some closely related species the ability to learn codes rather than just behaviors.

This conclusion might be strengthened if future research were to confirm observations of deliberate teaching of young apes by adults. Unfortunately, such observations are rare enough that to date they constitute only anecdotal evidence. Moreover, it is important to distinguish teaching of codes from teaching of behaviors. For example, King (1999) described several instances in which a primate mother encouraged an infant to cling or to crawl. This may constitute teaching, but what is being taught is a behavior, not a code.

The best example of which I am aware of the teaching of a code as opposed to a behavior is an observation by Boesch (1991:532) of a chimpanzee trying to crack nuts with a stone hammer:

> Ricci's daughter, 5-year-old Nina, tried to open nuts with the only available hammer, which was of an irregular shape. ... Eventually,

after 8 min of this struggle, Ricci joined her and Nina immediately
gave her the hammer. Then, with Nina sitting in front of her, Ricci, in
a very deliberate manner, slowly rotated the hammer into the best position with which to pound the nut effectively. As if to emphasize the
meaning of this movement, it took her a full minute to perform this
simple rotation. With Nina watching her, she then proceeded to use
the hammer to crack 10 nuts (of which Nina received six entire ker-
nels and a portion of the other four). Then Ricci left and Nina re-
sumed cracking. Now, by adopting the same hammer grip as her
mother, she succeeded in opening four nuts in 15 min. Although she
still had difficulties and regularly changed her posture (18 times), she
always maintained the hammer in the same position as did her
mother. ... In this example, the mother corrected an error in her
daughter's behaviour and Nina seemingly understood this perfectly,
since she continued to maintain the grip demonstrated to her.

According to this description, it appears that the mother was trying to
demonstrate a principle, not a behavior. If so, then she was trying to
transmit a code rather than to elicit a behavior. It should be noted, however, that this case is essentially a unique example. A few possible observations of the teaching of codes have been made among captive apes.
For example, Washoe, a chimpanzee trained in American Sign Language, is said to have molded her adopted infant's hands into the sign for
food (Fouts 1987:67; Fouts et al. 1982).

Overall, the data seem to indicate that an underlying neural capacity for learning coding from conspecifics predates the separation of
our lineage from that of the great apes. This is, however, only one part of
an ability to create coding through social interaction.

A few observations made in the course of ape language experiments
hint at a capacity in apes to create new linguistic codes. Various chimpanzees have created combinations of symbols to designate things for
which their human trainers had provided no signs. For example, after
tasting and spitting out a radish, the chimpanzee Lucy signed *cry hurt
food.* She also signed *candy drink* or *drink fruit* for watermelon (her vo-
cabulary did not include the symbols *water* and *melon*) (Fouts and Budd
1979:386-387). Miles (1990:535) reported that a language-trained
orangutan invented signs for, among other things, contact lenses and for
indicating that he would not use his teeth during play. Patterson
(1978:191) reported that a gorilla created signs for "bite," "stethoscope,"
"tickle," and "darn."

I leave it to primatologists and psychologists, who are better qualified than I am, to determine whether such observations do in fact demonstrate that apes are creating novel linguistic symbols. If they are indeed
doing so, then as soon as their human contacts understand these novel
signs, new codes have been created socially. This would mean that at
least some species of ape are capable of creating such codes. There is,

however, one caveat. These examples involve the creation of a new code through the social interactions not of two or more apes but between an ape and a human or humans. I am unaware of any example of an ape teaching a new symbol he or she has created to another ape, or of two apes agreeing on the meaning of a new symbol.

4.1.2. Behavior in the Wild

Although apes may be capable of creating simple coding through social interaction in captivity, it does not follow that they do so in the wild, or that such coding forms a part of their adaptation. The ethological literature, as far as I can tell, is almost devoid of any indication that they do, with one intriguing possible exception. Apes, like some other species, hunt cooperatively.

Although cooperative hunting may involve some coordination among the participants, it is not clear that this coordination is achieved through socially constructed coding. For example, chimpanzees often surround a potential victim, maneuvering so that all or most escape routes are cut off, adjusting their positions as the prey moves (Goodall 1986:286; Watts and Mitani 2002:13). This can probably be explained if each chimpanzee simply recognizes the potential escape routes, with no further coordination among the hunters.

A more complex encircling behavior has occasionally been observed among wolves and lions:

> Four lionesses walking down a road in single file see about ten gazelle by a riverine thicket 65 m away. While three wait, one lioness continues on alone, partly screened by acacia saplings. The gazelle spot her after 40 m and trot to one side. The lioness runs toward them, causing four to double back, two of which run directly in front of the three waiting lionesses. These rush, and one captures a subadult male after 40 m. (Schaller 1972:249)

Five wolves stalking a group of Canadian barren grounds caribou (*Rangifer tarandus*) that had entered a clump of spruce were observed following a similar strategy. One concealed itself on the path the caribou were following, while the other four circled behind the clump of spruce, spread out, and began to stalk toward the caribou. The hunt failed because of an apparent mistake by a yearling cub (Kelsall 1968:252, in ; Mech 1970:230).

Among humans, this kind of behavior would in most instances be guided by socially constructed coding. Such coding might be standard operating procedures known to all the hunters involved, careful planning of a game drive, or simply spur-of-the-moment agreement on how to proceed. However, it is not clear that such coding is *necessary* for this kind of coordinated behavior. For one thing, the observations are few enough that one should probably still consider the evidence to be anecdotal. However, even if the data point to a real pattern of behavior, there may be simpler explanations.

Let us analyze the behavior of the lions in the preceding example. When they sight a herd of gazelle, they carry out two complementary, mutually interdependent tasks. Three lions remain hiding in ambush while the fourth circles the herd and drives it toward them. In order for this to happen on a regular basis, the ambushers must know that the fourth lion will circle the herd and drive it toward them. The circling lion must know that the others will remain behind in ambush. What is necessary is simply that each individual be able to predict the *behavior* of the others, something that might have been learned by repeated observations during years of hunting together. It is not at all clear that any of the lions need consider what the others are thinking.

There is ethological and laboratory evidence (Cheney and Seyfarth 1990; Tomasello and Call 1997) that primates other than apes (and perhaps even some apes) are largely unaware that other individuals are "intentional" beings, that is, that their behavior is governed by mental states (Dennett 1987). It appears, for example, that a monkey's attitude when threatening a rival is not, "If I display, he will be afraid and move away." Such an attitude recognizes that the rival has a mind and mental states. Rather, the monkey apparently has a simpler view of the cause and effect involved: "If I display, he will move away."

If monkeys are incapable of recognizing mental states in other individuals, then the same is almost certainly true of lions and wolves. Since this recognition is essential for the kind of communication involved in creating emergent coding, it follows that the kind of cooperative hunting found among lions and wolves does not involve emergent coding. Thus, even in a species such as chimpanzees, which apparently do have this sort of awareness, such behavior cannot, in and of itself, constitute evidence of social creation of coding.

A group of researchers has suggested that wild bonobos (pygmy chimpanzees) in Zaïre (Democratic Republic of the Congo) deliberately leave messages for other members of the group (Savage-Rumbaugh et al. 1996). I leave it to primatologists to evaluate the validity of the evidence, but if we accept provisionally the researchers' interpretation of what the bonobos are doing, it will shed some analytic light on what is involved in the social creation of coding.

The bonobos travel from one food site to another through dense vegetation. They often split up, leaving a food site in separate groups that follow the same route to another site even when multiple routes are possible and when the preceding group is out of sight. Savage-Rumbaugh and her colleagues believe that those in the lead leave purposeful indications of their direction of travel, either by deliberately breaking or flattening vegetation or by leaving visible footprints. Such behavior would certainly point to something very close to the creation and use of emergent codes. If we assume that the bonobos are doing what Savage-Rumbaugh and her group think they may be doing, then clearly they are using signs to communicate information to others. This means that they are aware that other individuals are able to interpret the flattened vegetation and footprints as evidence of the passage and direction of travel of bonobos.

Even so, one thing is missing. Such signs would not depend on social interaction for their creation. Broken and flattened vegetation, footprints, and the like are what Peirce (1932 [1960]) called indexes, because they are naturally associated with the passage of bonobos. An index involves recognizing an existing correlation between two or more phenomena that one observes in one's environment. Because this link is observable, it is something that the individual bonobo can learn on its own. The index is a code – it is learned and it can be used to govern behavior – but it is a privately created code. If indexical signs such as flattened vegetation are used for communication, then only individual codes are used. Although the physical *signs* are created deliberately, no new *codes* are created through the interactions of multiple individuals. If the bonobos created what Peirce would call symbols – signs that carry meaning by arbitrary and agreed-upon convention, then that convention would be socially created.

The distinction becomes clear if we consider a sign (symbol) created by the humans whose job it was to track the bonobos: "three fruits placed in a line on the ground meant that bonobos nested nearby" (Savage-Rumbaugh et al. 1996:179). The line of three fruits had no natural connection to its meaning – it had to be constructed by convention among the trackers. By contrast, either a human or a bonobo can learn the meaning of flattened vegetation simply by observing what happens naturally when bonobos move through vegetation.

4.1.3. Implications of the Primate Evidence

At the beginning of this section, I argued that creating coding through social interaction, or the ability to do so, need not be a unitary, all-or-nothing matter, and that the evolution of the social construction of coding could be broken down into three separate phenomena:

- *The cognitive ability to learn codes (rather than just behaviors) from conspecifics.* The ability of at least some apes to learn

codes (in the form of symbols) from humans seems to be generally accepted by primatologists. The evidence for their learning
codes (as opposed to behaviors) from other apes appears to be
less well established.

- *The cognitive and social ability to create codes through social
  interaction.* If observations of laboratory apes creating new signs
  have been correctly interpreted, then apes do have the cognitive
  abilities required to create new codes socially, but this has occurred only when apes were interacting with humans, not among
  themselves.
- *The actual use of socially constructed codes in the wild as part of
  the adaptive repertoire of a nonhuman species.* I know of no
  good evidence that any species other than humans actually creates and uses such codes in the wild.

All this implies that we share with at least some other species many
of the basic mental and social abilities that underlie the creation and use
of socially constructed coding. Thus at least some of the foundations for
emergent coding had evolved before our lineage separated from that of
the pongids.

However, the matter is complex and merits further investigation.
Tomasello and his colleagues (Tomasello et al. in press) have hypothesized that the crucial difference between humans and other primates involves intentional abilities:

> Our proposal for this "small difference that made a big difference" is
> an adaptation for participating in collaborative activities involving
> shared [intentionality] – which requires selection during human evo-
> lution for powerful skills of intention-reading as well as for a motiva-
> tion to share psychological states with others. In ontogeny these two
> components – the understanding of intentional action and the motiva-
> tion to share psychological states with others – intermingle from the
> beginning to produce a unique development pathway for human cul-
> tural cognition, involving unique forms of social engagement, sym-
> bolic communication, and cognitive representation. Dialogic cogni-
> tive representations, as we have called them, enable older children to
> participate fully in the social-institutional-collective reality that is
> human cognition.

In spite of the abilities of apes, one thing must be kept in mind.
There is a stark contrast between humans and apes in terms of the role
socially created coding plays in our adaptations and in our daily lives. As
Gibson (2002:325) put it, "although several primatologists have stated
that chimpanzees and great apes or other primates possess culture, these
claims rest on definitions of culture as socially learned behaviors, rather

than as symbolic systems." The root cause may not be a sharply defined qualitative difference in mental abilities, but whatever the neurological correlates may be, the fact is that our ways of life are very different in this regard. This has two implications. First, tracing the evolution of socially created coding as an adaptation is of great importance for understanding *hominin* evolution. Second, evidence must be sought not among living primates but in the historical sciences of human paleontology and archaeology.

4.2. THE SKELETAL EVIDENCE FOR LANGUAGE

Human paleontologists must rely on very incomplete evidence from the past. Basically, all that is left to them of the anatomy and physiology of prehistoric hominid populations are bones and teeth. These are often discouragingly fragmentary, and for the period in which we are interested they are often so few as to raise serious questions about population variability and about the statistical validity of paleontological interpretations. However, two sets of data are available that are indirectly relevant to the origins of socially created coding. First, artificial or natural casts of the interiors of the braincases of sufficiently well-preserved skulls provide some information about the anatomy of the brain. Second, the base of the cranium, the palate, the mandible, and, in one single case, a hyoid provide evidence, albeit incomplete, of the anatomy of the vocal tract.

I say that these two kinds of evidence have only an indirect bearing on the origins of socially created coding. The reason is that, at least to date, they have been applied to understanding the origins of language rather than of the larger phenomenon of socially created coding. Language is one form of socially created coding, but the social creation of coding is possible, at least theoretically, without language. Thus, tracing the origins of language is tracing the origins of only one subset of socially created coding. Moreover, if skeletal evidence for language appears only after language becomes important enough to apply significant selective pressure on anatomy, then the paleontological record provides only a *terminus ante quem* for the origins of socially created coding – it cannot have evolved any later than the skeletal evidence for language, but it is possible, indeed extremely probable, that it evolved earlier.

4.2.1. Vocal Tract Anatomy

There has been considerable controversy over the evolution of the human vocal tract, a debate that for various reasons has concentrated primarily on the position of the larynx in Neanderthals. The position of the larynx is of importance because it affects the production and use of formant frequencies, which are characteristic of the speech of fully modern *Homo sapiens.*

4.2.1.1. Background

(The following explanation of this phenomenon is based on that of Lieberman [1984], which the reader should consult for a fuller account.)

To produce a vowel, the vocal cords in the larynx create a sound consisting of a fundamental frequency and its harmonics. This sound is filtered through the resonating cavity of the vocal tract above the larynx. A speaker controls the shape of the resonating cavity by the position of his or her tongue, lips, and jaw. Changes in the shape of this resonating cavity change the degree to which each frequency is permitted through the filter (Figure 4.1). Peaks in the frequencies permitted to pass through the filters are called formant frequencies. Each vowel is characterized by a distinct pattern of formant frequencies. The ear and brain of the human hearer recognize patterns of formant frequencies that are linked causally to specific configurations of the movable parts of the vocal tract. Other factors, such as nasalization, may be involved as well, but formant frequencies are of elemental importance.

The vocal communications of other primate species may also involve formant frequencies (Andrew 1976; Lieberman 1994), but there are good reasons to believe that humans have evolved anatomical and neural specializations for efficient speech. For example, consonants in which the flow of air is completely interrupted (such as [b], [d], and [g]) are called stops. At the point where air flow is cut off, a stop is entirely silent. Because different stops cannot be distinguished on the basis of silence, the hearer recognizes them by the change or deformation of the formant frequency patterns of the vowels immediately preceding or following them (Figure 4.1c). As the mouth changes from the position of the preceding vowel to that of the stop, and then to that of the following vowel, the pattern of the formant-frequency filter changes, and it is these changes that identify the stop for the hearer.

The effect of the same stop on the formant frequencies of different vowels is different. The consonant [b], created by sealing the mouth with both lips, is acoustically different for each vowel with which it is associated – yet our brains are designed to hear all of these as the same stop. The stop cannot be perceived except in conjunction with a vowel or vowels, and formant frequencies are essential to the process of perceiving both vowels and stops. To me, the fact that our brains lump dissimilar changes in formant frequencies on the basis of the way they are produced anatomically indicates that this is a specialization particular to speech (see Schoenemann 1997 for a different perspective).

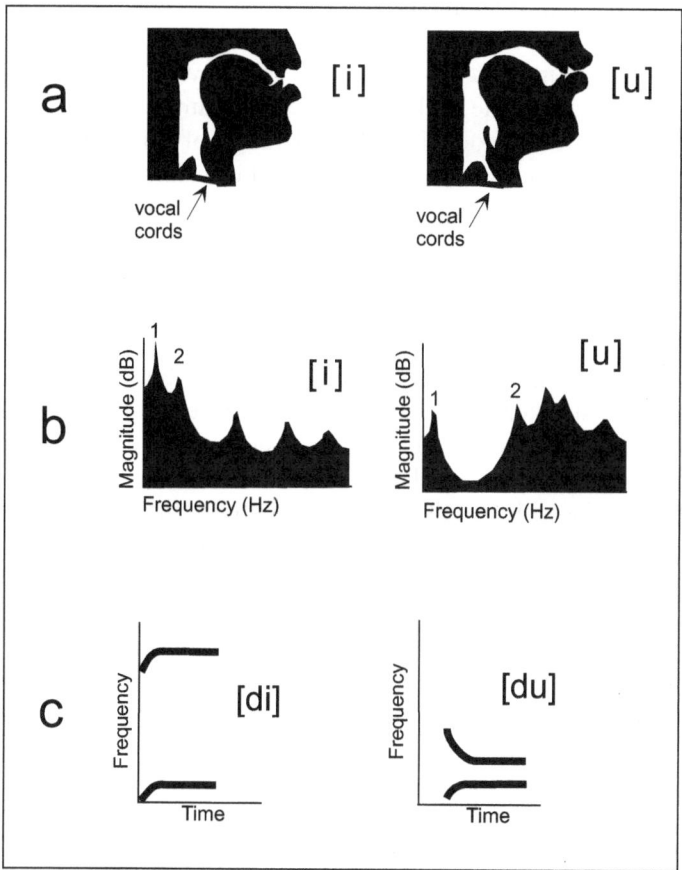

Figure 4.1. (a) The shape of the supralaryngeal vocal tract when making the vowels [i] and [u]. This resonating cavity acts as a filter for sound produced by the vocal cords. (b) The filter imposed on this sound. (The graph shows how white noise would be filtered, but the vocal cords actually produce sound only at a fundamental frequency and its harmonics.) (c) The effect of the stop [d] on the vowels [i] and [u]. As the tongue moves after the stop is released, the shape of the resonating chamber changes, and as a result the formant frequencies of the filter change. Although the pattern of changes is different for each vowel, the human brain interprets both patterns as the same consonant. (After Lieberman 1984, figure 7-2, and Lieberman et al. 1972, figure 8)

For our purposes, three important arguments have been made concerning the evolution of this system of speech production and perception (Lieberman 1984; Lieberman 1989; Lieberman 1991; 2001; Lieberman and Crelin 1971; Lieberman et al. 1972). First, the full range of vowels used in present-day languages can be produced only if the larynx is located much lower in the throat than is the case for either human infants or nonhuman primates. In these, the larynx is located high in the throat, so that almost the entire supralaryngeal vocal tract consists of the oral and nasal cavities. In present-day adult humans, the portion below the oral cavity is approximately as long as the oral cavity itself and is oriented at a sharp angle to the oral cavity. Neither human infants nor apes appear to be capable of producing the full range of formant frequency patterns that adult humans can produce. This fact, along with computer simulations, indicates that the production of the full range of vowels requires a lowered larynx and an angled supralaryngeal vocal tract.

Second, one of the vowels that cannot be made without a lowered larynx is particularly important in present-day human speech. This is the vowel [i] (as in "bee"). When different individuals make this vowel, the formant frequencies are not the same in absolute terms but are determined by the size of the speaker's vocal tract. In order to recognize a vowel accurately, therefore, the hearer must have some way of determining the absolute size of the vocal tract. For example, in absolute terms, the [ɑ] (as in "bah") produced by adult males is essentially the same as the [ɔ] (as in "bawl") produced by adolescents. The hearer's brain will "automatically" distinguish between the two if, but only if, it has clues to the size of the vocal tract. By contrast, [i] is distinct from other vowels no matter what the size of the vocal tract and can therefore serve as a clue to the absolute size of the vocal tract. Having [i] in the repertoire of vowels would add to the efficiency of speech.

Third, lowering of the larynx has an adaptive cost. In human infants, the larynx is high enough that the epiglottis can form a seal with the soft palate; air can pass from the nasal cavity to the larynx through a passage that is completely sealed off from the pharynx, the passage through which food and drink must pass. Breathing and swallowing can occur simultaneously. When the larynx is lowered, this isolation of the larynx is impossible. During swallowing, solids or liquids pass around the larynx to the esophagus. The result is that there is always the danger, during swallowing, that food will enter the larynx rather than bypassing it, with the possibility that the airway will be clogged and asphyxiation will occur.

These three arguments have encountered little criticism from others, with two exceptions. First, Duchin (1990) argued that the portion of the vocal tract behind the oral cavity is essentially irrelevant to articulation, a claim effectively countered by Lieberman (2001; Lieberman et al. 1992). Second, Larson and Herring (1996) used videoflouroscopy to observe swallowing in pigs and ferrets, which, like apes, human infants, and most other mammals, have larynxes located high, close to the soft palate. Surprisingly, they found that swallowing was the same as in adult humans, with the epiglottis moving down and away from the soft palate, rather than forming a seal with it.[*] With these exceptions, most of the controversy has involved the reconstruction of the position of the larynx in fossil hominids.

4.2.1.2. Discussion

Before we move on to the topic of the larynx in fossil hominids, it is worthwhile to stop and consider, from the perspective of the evolution of socially created coding, exactly what is at stake.

- The descent of the larynx does *not* mark the first use of formant frequencies by primates.
- It does *not* mark the origin of socially created coding.
- It does *not* mark the beginnings of language.
- It does *not* mark the origins of articulate speech.

What it does indicate is that the full range of vowels and consonants used in present-day languages could have been available to the hominin in question, and that certain tools for efficient speech could have been available to facilitate recognizing the size of the vocal tract. (These points have been made explicitly by Lieberman and his colleagues (Lieberman 1976b; Lieberman et al. 1972:302; Lieberman et al. 1992:464), although they have not always been appreciated by their critics.)

Thus the debate concerns speech directly and socially created coding only indirectly. Indeed, it concerns even language per se only indirectly. Let me define language as a form of socially created coding by which symbolic semantic and syntactic conventions are used to communicate meaning; and let me define speech as the vocal production and aural perception of sounds used as a medium for language. Given these definitions, the use of language must predate specialized anatomical adaptations for fully modern speech. Because all primates use vocal sounds for communication, it follows that by the time the vocal tract evolved in order to facilitate *linguistic* communication, language must already have been a significant part of the adaptation of the population involved. If the adaptive costs of the changes involved were high (in terms of choking), then it follows that the adaptive importance of language, which these changes served, must also have been high.

Thus skeletal evidence for a low larynx is a *terminus ante quem* for the evolution of language as a significant part of hominid adaptation, and

---

[*] A study of chimpanzees by Nishimura et al. (2003) indicated that there was an earlier descent of the larynx, before the chimpanzee and human lineages separated. This involved a descent of the laryngeal skeleton relative to the hyoid, rather than of the hyoid itself, and it may have conferred and advantage in adult swallowing.

this in turn is a *terminus ante quem* for the evolution of socially created coding. The absence of a low larynx implies neither the absence of articulate speech or language nor the absence of socially created coding.

The claim that Neanderthals did not have a modern vocal tract with a low larynx (Lieberman and Crelin 1971; Lieberman et al. 1972; Lieberman et al. 1969) set off a heated debate within human paleontology. The claim was based on the Neanderthal cranium from La Chapelle-aux-Saints in France. Lieberman and Crelin determined the position of the hyoid by reconstructing the intersection of the geniohyoid muscle and the stylohyoid ligament as indicated by the inclination of the styloid processes, two projections from the base of the cranium. Lieberman (1984:294-296) also argued that the Neanderthal palate was so long that the larynx could not have been low enough for fully modern speech without being located below the neck in the chest.

These reconstructions have been attacked on anatomical grounds, on the basis of the specimen and cast used, and for other reasons (Burr 1976b; 1976a; Carlisle and Siegel 1974; Du Brul 1976; Falk 1975; Houghton 1993; LeMay 1975; Mann and Trinkaus 1973; Morris 1974; Wind 1989). All of these critiques have been countered (Lieberman 1973; 1976b; 1976a; 1989; 1991; 2001; Lieberman and Crelin 1974), and it appears that there is as yet no consensus among anatomists and paleontologists.

The discovery of a hyoid with a Neanderthal skeleton from Kebarah, Israel (Arensburg 1989; Arensburg et al. 1990; Arensburg et al. 1989) provided a new set of information but no agreement. The cranium of this specimen was missing, although the mandible was present. The team that discovered and described the skeleton argued that the hyoid was essentially modern in form (although the mandible was not), and that this implied a modern anatomical position. Others disagreed that Arensburg and his colleagues had demonstrated that the hyoid was in fact modern in its morphology (Laitman et al. 1990; Lieberman et al. 1989). Lieberman (1993) also argued that because the form of the hyoid does not change its shape in present-day humans during its descent in the growing child, the argument that a modern hyoid cannot be located high, as in infants, is not demonstrated. There is still little consensus about these data.

There has been somewhat less controversy over a different method of reconstructing the position of the larynx in fossil populations, even though the logic behind it is probably less direct. Laitman, Heimbuch, and Crelin (1978) had noted a correlation between the form of the cranial base and the position of the larynx in apes, modern human infants, and modern human adults. The first two have a flatter cranial base; human adults have a more flexed cranial base (Figure 4.2). Although the functional reasons for this relationship were not specified, the increased flexion of the cranial base during the development of human children is correlated with the descent of the larynx (Crelin 1987).

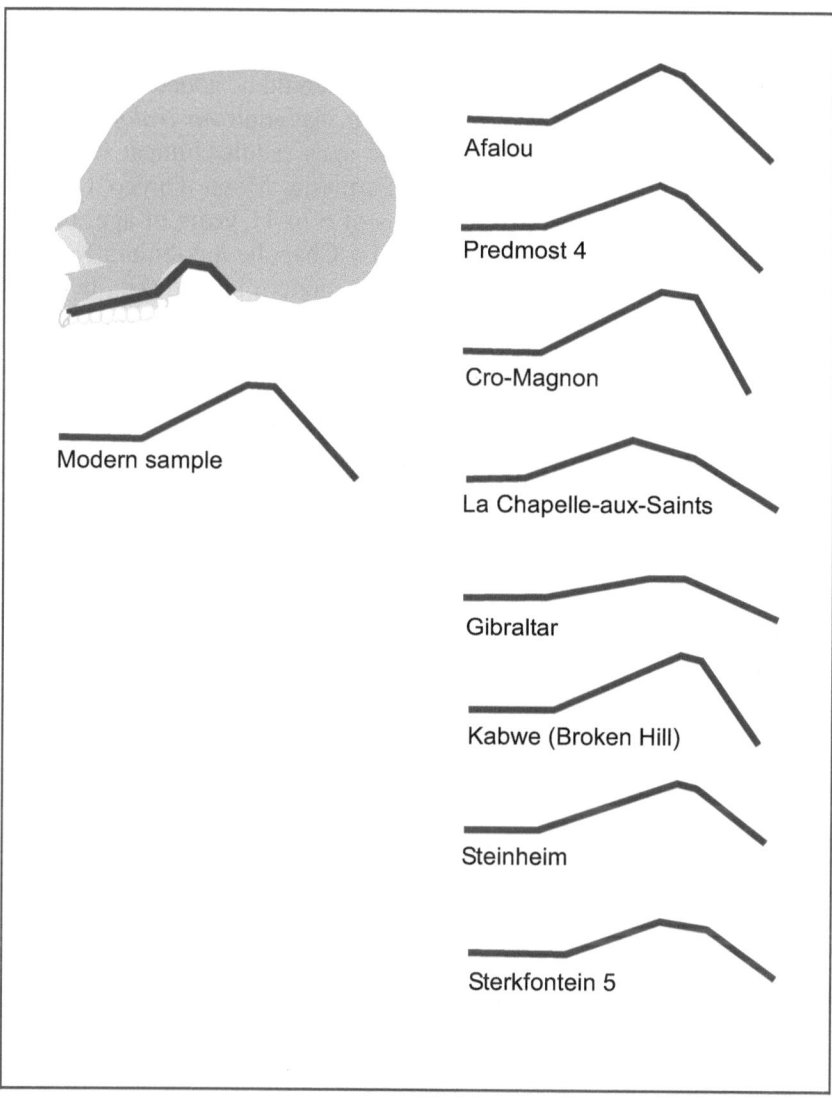

FIigure 4.2. Laitman, Heimbuch, and Crelin (1978; Laitman et al. 1979) determined basicranial flexion by means of lines drawn between landmarks on the base of the skull (prosthion, staphylion, hormion, sphenobasion, and endobasion). These were measured on modern humans and on various fossil skulls (arranged in chronological order from bottom to top). (Adapted from Laitman, Heimbach and Crelin 1978, figure 5.)

On this basis, the researchers used basicranial flexion as a way of estimating where various fossil hominids fell in comparison with both apes and modern human adults (Laitman et al. 1979). They based their study on a discriminant function analysis of 228 modern primate skulls, including humans, in five different categories of dental development (from prior to eruption of the deciduous dentition to fully adult). This provided a scatter plot to which the data from the fossils could be added.

The results of the analysis are interesting. An adult *Australopithecus* from Sterkfontein (Sts5) falls close to the apes (adult chimpanzee and young gorilla). Some Neanderthals (La Ferrassie, Monte Circeo 1, and Saccopastore 2) resemble modern humans of 6 to 11 years of age, indicating partial descent of the larynx. The La Chapelle-aux-Saints Neanderthal, however, resembles a young stage 2 modern human (i.e., before complete eruption of the deciduous dentition). This is in line with Lieberman and Crelin's (1971) analysis of the La Chapelle specimen as having a high larynx. Anatomically modern humans from the later Upper Pleistocene and from the Holocene (Predmošt, Cro-Magnon, Afalou, Taforalt, and Ain Dokhara) fit the modern pattern.

From our perspective, however, the most interesting findings concern the fossil skulls from Steinheim in Germany and from Kabwe (formerly Broken Hill) in Zambia. Both are essentially modern in terms of basicranial flexion and therefore presumably in terms of upper respiratory anatomy and position of larynx. These specimens have been variously classified as archaic *Homo sapiens, H. heidelbergensis, H. rhodesiensis,* and ancestral *H. neanderthalensis;* both are probably older than any of the classic Neanderthals in the sample.

The use of basicranial flexion to reconstruct larynx position has received fewer criticisms than the reconstructions of Lieberman, Crelin and others, although Arensburg, Shepartz et al. (1990) argued that the relationship between basicranial flexion and larynx position had not been demonstrated. Houghton (1993) argued that basicranial flexion should not be measured using both the braincase (sphenobasion and basion) and the bony palate (prosthion and staphylion), which are two separate functional systems. However, Laitman and his colleagues intentionally chose to define basicranial flexion in this way, because it changed in conjunction with the descent of the larynx during human development and because they believed it was closely associated with the anatomy of the upper vocal tract.[*] Houghton also questioned the measurability of sphenobasion and the use of the La Chapelle specimen.

---

[*] Lieberman and McCarthy (1999) argued that during growth and development in present-day human children, basicranial flexion and descent of the larynx are not correlated. However, they measured basicranial flexion internally, including the braincase but excluding the palate from their definition. Thus "basicranial flexion" means some-

The base of the skull is, of course, involved in much more than the upper respiratory system, and other hypotheses have been proposed for the evolution of basicranial flexion, involving brain size, facial architecture, and mastication (Strait 1999). Thus, even if basicranial flexion is correlated with a lowering of the larynx, it is logically possible that basicranial flexion evolved in response to selective pressures other than speech.

If so, and if basicranial flexion was a preadaptation or exaptation rather than an adaptation for fully modern speech (in terms of the range of vowels available), then its appearance in the fossil record constitutes a *terminus post quem* for fully modern speech. Because the latter is simply a *terminus ante quem* for language and socially created coding, basicranial flexion would in this case tell us almost nothing about the origins of the latter phenomena. Socially created coding must have become a significant part of hominin adaptation before fully modern speech, but fully modern speech could have evolved after basicranial flexion.

It would be foolhardy for an archaeologist to try to evaluate the arguments and the evidence I have cited when anatomists and paleontologists are not in agreement. I can, however, summarize what I see as the implications of the various claims and counterclaims for our ability to date the origins of socially created coding. Let me begin with the following assumptions:

- Language is a form of socially created coding. It is possible, perhaps even likely, that simple socially created coding other than language appeared earlier, but the use of language would have permitted much more complex and effective coding to have been created through social interaction. Thus the evolution of socially created coding as a significant part of hominin adaptation was probably to some degree correlated with the evolution of language.
- Evolution of anatomical structures or configurations that enhanced the effectiveness of speech would have occurred only after language (and therefore socially created coding in general) became a significant part of hominid adaptation. This argument

---

thing different as defined by Lieberman and McCarthy and as defined by Laitman, Heimbuch, and Crelin. Ross and Henneberg (1995) used internal measurements of basicranial flexion to compare fossil with modern specimens. Both a *Homo erectus* from Olduvai Gorge (OH9) and three *Australopithecus* specimens (St5 from Sterkfontein and MLD 37/38 from Makapansgat) showed the same degree of basicranial flexion as modern humans. They noted that since Laitman, Heimbuch, and Crelin (1979) found major differences between St5 and modern humans, this indicated that internal and external measures of basicranial flexion are not necessarily correlated. This implies that Lieberman and McCarthy's findings are probably irrelevant to the question at hand.

is stronger if the evolution of such structures entailed an adaptive cost (the possibility of choking when swallowing) and weaker if in fact we swallow in essentially the same way as other mammals.

- In any case, the evolution of anatomical structures or configurations that enhanced the effectiveness of speech serves as a *terminus ante quem* for language and socially created coding.

Given these assumptions, whether or not the fossil evidence for the vocal tract is relevant to our problem depends on our fulfilling four conditions:

1. Identifying valid anatomical criteria for reconstructing the location of the larynx using skeletal material
2. Ascertaining that these anatomical features are an adaptation rather than a preadaptation or exaptation for a modern range of speech sounds
3. Applying these criteria to a range of fossil material, not just to Neanderthals
4. Obtaining reliable dates for the earliest fossils showing evidence of a modern position for the larynx

In the course of the debate cited earlier, two claims were made for meeting the first criterion. Lieberman and Crelin used the angle of the styloid process, and Laitman, Heimbuch, and Crelin used basicranial flexion. These two claims differ in the structure of their logic. The first is a straightforward anatomical argument – the angle of the styloid process indicates the angle of the stylohyoid muscle and therefore the direction to the hyoid. The latter is based on a correlation. Among present-day primates, only adult humans have a low larynx. They also have a high degree of basicranial flexion (as defined and measured by Laitman et al.). However, no functional argument has been presented to explain exactly how basicranial flexion is *causally* related to the descent of the larynx.

Not being an anatomist, I will not attempt to evaluate the relative merits of Lieberman and Crelin's reconstruction of the position of the La Chapelle vocal tract, or criticisms of that reconstruction. It seems unfortunate to me, however, that the debate has been so focused on the speech of Neanderthals, rather than on the fossil record as a whole. Lieberman and his colleagues have not applied their methods of reconstruction to other fossils but have been content to cite the work of Laitman and his colleagues. Their critics have, for the most part, presented arguments disputing Lieberman and Crelin's analysis of La Chapelle but have dropped the matter there, rather than going on to investigate the origins of speech in a more general manner. For example, both Arensburg, Schepartz et al. (1990:141) and Houghton (1993:141-143) cite the anatomy of the Neanderthal mandible in arguing for an essentially modern vocal tract in Neanderthals, but they do not apply their methods to the mandibles of other fossils. The result is that the arguments on either side, as they stand today, may tell us something about the speech anatomy of the La Chapelle specimen but are of little use for dating the origins of fully modern speech. They will remain irrelevant until they are made applicable to and have been applied to a wide range of fossils, as demanded by the third criterion I listed.

Laitman and his colleagues have defined a criterion for recognizing a modern vocal tract anatomy that is applicable to any fossils that are sufficiently well preserved, and they have in fact applied them to a wide range of fossils. The problem with their methodology lies rather with the first two criteria. If a causal relationship could be established between basicranial flexion, a low larynx, and adaptation for modern speech, then both of these criteria would be met. Until they or someone else is able to do so, however, it seems to me that however valid their anatomical arguments may be, their results, though suggestive, are not wholly conclusive. This is because basicranial flexion, even if it is correlated with the descent of the larynx during development, may represent a exaptation rather than an adaptation for fully modern speech.

The final problem with the present state of our knowledge is archaeological or geological rather than biological in nature. If we accept Laitman, Heimbuch, and Crelin's analysis of fossil hominins, then the oldest specimens with an essentially fully modern vocal tract are those from Kabwe and Steinheim. Unfortunately, neither is as well dated as one might wish.

The Kabwe specimen was discovered in 1921 during mining operations, rather than by archaeological excavation, and subsequent mining has completely destroyed the sediments from which it came. As a result, the provenience of the specimen is somewhat suspect. On the basis of faunal remains and stone tools from the vicinity, the skull can probably be dated to the late Middle Pleistocene (Klein 1973; Rightmire 1981; 1984), meaning that it is more than 127,000 years old. Klein (1994) and Rightmire (1998) have subsequently suggested a date of 400,000–700,000 years, and McBrearty and Brooks (2000) make a much older estimate (780,000 to 1.33 million years). Unfortunately, the association of the skull with either the stone tools or the faunal remains is uncertain.

The Steinheim specimen was discovered under somewhat more favorable circumstances, but its dating is not entirely certain either (Stringer 1981). It has generally been considered to date to the Middle Pleistocene (Adam 1985; Howell 1960; Klein 1999:306; Schwartz and Tattersall 2002:347).

If this dating is correct, then the origins of language and socially created coding lie well back in the Middle Pleistocene and probably even earlier. On the other hand, if, as Larson and Herring's (1996) videoflouroscopy implies, no major adaptive cost is involved in lowering the larynx, then the origins of language may lie closer to the anatomical changes reflected in the base of the skull. In other words, language must already have been a significant part of hominid adaptation for the benefits of more efficient speech to have offset the danger of choking when swallowing; but it could have been of less significance if there was no such cost. In either case, language and socially created coding date at least to the Middle Pleistocene.

If, however, basicranial flexion has nothing to do with the position of the larynx, or if basicranial flexion is an exaptation permitting the eventual lowering of the larynx, then we must turn elsewhere for evidence for the origins of modern human speech. The controversy surrounding Lieberman and Crelin's reconstruction of the La Chapelle Neanderthal's upper vocal tract applies directly only to the speech capabilities of Neanderthals. If Lieberman and Crelin's use of styloid processes to locate the position of the hyoid and larynx is valid, then it must still be applied to other fossils before a good *terminus ante quem* for language can be established.

If the lowering of the larynx itself is an exaptation rather than an adaptation for producing the modern range of speech sounds, then no matter how well paleontologists are able to reconstruct its position in fossil hominins, it will tell us nothing about the origins of language. The lowering of the larynx will be simply a *terminus post quem* for a *terminus ante quem* (specialization for modern speech) which itself cannot be dated. (Fitch (2000) has suggested one possible adaptive value, other than human speech, of a longer supralaryngeal tract: to enhance a person or animal's apparent size.)

At present, only two things can be said with certainty: there is still considerable disagreement among experts, and there is still a great deal of work to be done before the fossil evidence for modern human speech specializations can be applied with confidence to determining the date, even a *terminus ante quem,* for the origins of language and socially created coding.

4.2.1.3. Innervation of the Tongue and Thorax

4.2.1.3a. Hypoglossal Canal Size. The hypoglossal canal, an opening through the occipital bone in the base of the skull, carries all but one of the motor nerves of the tongue. Hypothesizing that the size of the canal would be correlated with the number of motor units in the tongue, and therefore with control of tongue shape during speech, Kay, Cartmill, and Balow (1998) compared this feature of modern humans with those of apes and of fossil hominins. Their results showed that the sizes of the hypoglossal canal in chimpanzees, bonobos, gorillas, and orangutans were all smaller than those of present-day humans, although the ranges overlapped, especially between gorillas and humans. However, the ratio of canal size to the size of the oral cavity was much higher in humans than in gorillas, implying that the human tongue is much more richly innervated.

For fossil specimens, they found that two Middle Pleistocene specimens (Kabwe and Swanscombe), two Neanderthals (La Chapelle-aux-Saints and La Ferrassie 1), and Skhul V fell well within the range of present-day *Homo sapiens,* whereas five *Australopithecus* specimens from Sterkfontein fell well outside that range, resembling living apes.

However, an attempt by DeGusta, Gilbert, and Turner (1999) to replicate these results failed. Hypoglossal canal size, corrected for palate size, overlapped drastically among present-day humans, apes, and Old and New World monkeys. *Australopithecus, Paranthropus,* and Neanderthal measurements fell well within the range of both modern humans and apes. Moreover, they found no significant correlation in cadavers between hypoglossal canal size and number of axons, indicating that innervation of the tongue is not correlated with hypoglossal canal size.

4.2.1.3b. Innervation of the Intercostal Muscles. MacLarnon and Hewitt (MacLarnon 1993; MacLarnon and Hewitt 1999) showed that the skeleton of a *Homo ergaster* or *Homo erectus* boy was characterized by a smaller thoracic vertebral canal (allometrically scaled to estimated body weight) than were living or fossil *Homo sapiens* or Neanderthals. The specimen, KNM-ER 15000, was from Nariokotome in Ethiopia, dated to just slightly more than 1.5 million years ago (Brown and McDougall 1993). KNM-ER 15000, as well as three *Australopithecus* specimens (AL 288-1, Sts 14, and Stw 431) fell close to the line for all primates; the *H. sapiens* and Neanderthal specimens fell at the very high end of the distribution.

MacLarnon and Hewitt rejected most possible explanations for this observation – the demands placed on the muscles of the thorax and abdomen by bipedalism, by efficient throwing of weapons, or by giving birth to large-headed infants – on the grounds that these would have evolved earlier than the date of KNM-ER 15000. They concluded that the reason for the expansion of the thoracic vertebral canal in the later hominins was to provide increased breath control. They reviewed the evidence that in living humans, speech is enhanced by changes in the breathing pattern that permit long periods when an even subglottal air pressure is maintained, broken by quick inhalations. This requires much finer control of the thoracic and abdominal muscles than is required for normal breathing. They concluded that *Homo ergaster/Homo erectus,* like *Australopithecus,* had not evolved the fine level of breath control involved in modern human speech.

Latimer and Ohman (2001) argued that KNM-ER 15000 suffered from numerous pathologies of the axial skeleton, including the vertebral column, and that "in view of these observations, suggestions regarding the biology and behavior of *H. erectus* that are founded upon the morphology of the axial skeleton must be carefully reexamined in the light of the described pathology." In addition, Meyer (2003) disputed MacLarnon and Hewitt's claim that the intercostal muscles (rather than the diaphragm, which is innervated from the cervical spine) are of major importance in breath control during speech.

Even if we ignore these criticisms, MacLarnon and Hewitt's findings, though of considerable interest in other contexts, do little to help us date the earliest language and the earliest socially created coding. If KNM-ER 15000 lacked the necessary breath control, it represents a *terminus post quem* for the evolution of modern speech, but this in turn is only a *terminus ante quem* for language. A *terminus post quem* for a *terminus ante quem* provides no date at all.

4.2.2. Cranial Endocasts

In studying the evolution of the brain, and its implications for the origins of language, human paleontologists and paleoneurologists must work from the anatomy, not of the brains themselves, but of the bony braincases that originally contained them. There are a number of ways in which neuroanatomists can study the relationship between braincase anatomy and language. Of particular importance for paleoneurology are the relationships between the patterns of folding of the cerebral cortex. Folding produces bulges (gyri or convolutions) separated by grooves (sulci or fissures). The patterns of folding divide the cerebrum into gross areas (frontal, parietal, and occipital lobes, prefrontal cortex, etc.) that have functional significance. Folding also provides landmarks for locating cytoarchitectonic areas (Brodmann's areas) that cannot be observed directly on an endocast but that have neurofunctional significance. However, reconstructing the form of the living brain is difficult, because layers of tissue and fluid intervene between the brain and the bone of the braincase.

Paleoneurologists have concentrated on four anatomical traits in conjunction with the evolution of language: the overall size of the brain, the reorganization of the brain in terms of differential increases in the sizes of different parts, the evolution of specific anatomical landmarks known to be involved with language in living humans, and asymmetries of different parts of the brain.

4.2.2.1. Brain Size

Certainly one of the most striking things about human evolution is the great increase in brain size relative to other primates, and few would deny that there is some relationship between this increase and the evolution of human intelligence, including language ability. Given the fragmentary or distorted condition of most fossil crania, brain volume is often difficult to measure. Moreover, because a natural allometric relationship exists between brain size and body size across all mammalian species (larger bodies require larger brains), encephalization is usually measured after adjusting for this relationship (Falk 1980a; Jerison 1982; Rightmire 2004:115-118). For fossil hominids, body size is even more difficult to estimate than cranial capacity. Nevertheless, these practical difficulties can be overcome.

The theoretical issues are probably more serious. In 1948, Keith (1948) estimated that a brain of at least 750–800 cc was necessary for language. Since then, probably because of a greater appreciation of the natural variation in human brain size, the idea that brain size alone can be taken as an indicator of the presence or absence of language has largely fallen out of favor (Falk 1980a; Holloway 1966; 1979; Schepartz 1993).

The most notable exception is Jerison (1976), who estimated the brain volume dedicated to language in living humans. For fossil hominids, he could then calculate how much brain volume was accounted for by allometry and how much was left over for language. However, because his figures for living humans related to fully developed modern language, he was unable to know with any precision how much brain volume would have been required for simpler proto-languages. Overall, his studies led him to conclude that *Australopithecus* had too little cranial capacity for language, that *Homo habilis* was questionable, and that *Homo erectus* was "borderline competent." (Jerison did not specify which specimens he used in his analysis as representative of each taxon.)

Of course, his analysis depended on the fact that he was "inclined to consider the evolution of language and language-related cognitive capacities as the source of essentially all of the advance from the pongid to the sapient hominid grade" (1976:377). If this were not the case, then he would have no way of knowing what amount of brain increase was due to the evolution of language and what amount was dedicated to the development of other forms of human intelligence – technology, planning depth, and so forth.

Deacon (1992; 1997) argued that brain size was useful as an index to language, but he did so on the grounds that as the size of the brain increased, the relative sizes of different parts of the brain changed, as did the connections among different parts of the brain. This is a more complex argument than Jerison's. In it, brain size is causally related to brain reorganization, and selection for symbolic and linguistic ability is seen as having both driven and resulted from this reorganization.

4.2.2.2. Brain Reorganization

Holloway (1966; 1972; 1983a) argued that reorganization of the brain was as important as the increase in overall size. By reorganization, he meant differential increases in various parts of the brain during the course of evolution, differences that could not be explained allometrically. The study of this reorganization has been a major focus of paleoneurology.

Probably the most controversial point has involved the relative sizes of two portions of the cortex, the parietal association cortex and the primary visual cortex. They are separated by a small sulcus, the lunate sulcus, whose position is potentially visible on the endocranium. In humans, when it is present, the lunate sulcus is in a more posterior position than it is in apes. Holloway (1972; 1975; 1976; 1981a; 1983c; 1983a; 1983b; Holloway et al. 2004; Holloway et al. 2003; Holloway and Kimbel 1986) has argued that the lunate sulcus in *Australopithecus* (*Paranthropus* and *Australopithecus*) is in a posterior position, indicating that the reorganization of the brain toward a human pattern began early, even before the accelerated increase in size that commenced with the genus *Homo*. This argument has been challenged by Falk on the basis of her reading of *Australopithecus* endocasts (Falk 1980a; 1985; 1986).

Holloway (1972; 1973; 1975; 1976; Holloway et al. 2004) also found differences, relative to apes, in the morphology of both the temporal lobes and the anterior frontal lobes of *Australopithecus* (*Australopithecus* and *Paranthropus*) and early *Homo*. Falk (1983) argued that the KNM-ER 1805 *Australopithecus* from Koobi Fora, Kenya, differed from the KNM-ER 1470 *Homo habilis* (or *Homo rudolfensis*) in the form of the frontal lobe. Falk, Redmond et al. (1999) found *Australopithecus* more similar to *Homo* than to *Paranthropus* on a number of traits, including the forms of the frontal and temporal lobes.

It would appear, then, that a reorganization of the brain was under way even before the appearance of the genus *Homo,* even if there is some disagreement about the details. However, with *Homo,* the reorganization was clearly established along human lines, and the brain was expanding more rapidly relative to body size than it had in *Australopithecus* and *Paranthropus.*

It should be noted that Bruner, Manzi, and Arsuaga (2003), on the basis of principle components analyses of endocast measurements, argued that the expansion of the brain in Neanderthals followed an allometric trend shared with earlier hominins. *Homo sapiens,* by contrast, followed a different allometric path that involved greater expansion of the parietal area. They did not attempt to explain exactly what this meant in terms of brain function.

4.2.2.3. "Language Areas"

Neuroanatomists have been able to locate areas of the brain that are involved in language production and comprehension on the basis of language pathologies in patients with localized brain damage and, more recently, on the basis of MRI scans of subjects performing linguistic tasks. Observations on fossil endocasts relevant to two of these areas have been cited as evidence for the origins of language. Broca's area, which involves Brodmann's areas 44 and 45, is located on the inferior or third frontal convolution. Wernicke's area is harder to delineate than Broca's area, but it includes parts of the parietal and temporal lobes near the posterior end of the Sylvian fissure. Both Broca's and Wernicke's areas play important roles in language production and comprehension and are usually located on the left side of the brain in right-handed people.

Holloway (1972; 1975; 1983b; Holloway et al. 2004) found at least some evidence for a developed Broca's area in endocasts of *Australopithecus* (*Australopithecus* and *Paranthropus*) as well as *Homo habilis* (*H. rudolfensis*) and *Homo erectus* (*H. ergaster*). Falk (1983) considered that the KNM-ER 1470 *Homo habilis* (*H. rudolfensis*) had a humanlike pattern in this area, although the same was not true for KNM-ER 1805 *Australopithecus* (or *Homo habilis* [Wood 1991]). Tobias (1987; 1991) found evidence of Broca's area and of an enlarged Wernicke's area in *Homo habilis.*

Tobias (1987) observed that the supramarginal and angular gyri (Brodmann's areas 39 and 40) were well developed in *Homo habilis* endocasts from Olduvai Gorge. This area, which lies between Wernicke's and Broca's areas, is generally considered to play a key role in language (Geschwind 1970). In addition, in all *Homo habilis* specimens in which it was observable, the superior parietal lobule was larger on the left than on the right side. This asymmetry distinguished *Homo habilis* from *Australopithecus.*

4.2.2.4. Lateral Asymmetries

There are lateral asymmetries in the functions of the human brain, due both to handedness and to language. Not surprisingly, anatomical asymmetries exist in the human brain as well. A number of anatomical asymmetries can be found in the brains of other primates, too, notably the great apes (Holloway and de la Coste-Lareymondie 1982; LeMay 1975; 1976; 1985; 1982; Semendeferi 2001). One pattern of asymmetry

– right frontal, left occipital petalia – is especially characteristic of humans. Our species exhibits a repeated pattern in which the right frontal (anterior) pole of the brain is larger than the left, while in the same individual the opposite is true of the occipital (posterior) pole. Although either or both of these asymmetries may be observed in individuals of other species, the repeated pattern of both asymmetries combined in multiple individuals is absent.

Skeletal evidence of, or at least evidence compatible with, the presence of right frontal, left occipital petalia has been reported for Neanderthals (Holloway 1985; Holloway et al. 2004; Holloway and de la Coste-Lareymondie 1982; LeMay 1976), for *Homo erectus* and *Homo ergaster* (Holloway 1981b; 1983b; Holloway et al. 2004; Holloway and de la Coste-Lareymondie 1982; LeMay 1976), for early *Homo* (Holloway 1983b; Holloway et al. 2004; Holloway and de la Coste-Lareymondie 1982), and for *Australopithecus* and *Paranthropus* (Holloway et al. 2004).

Unfortunately for our ability to date the origins of socially created coding, no causal link between this petalial pattern and language has yet been clearly demonstrated. The problem is that both language and handedness (which is probably related to tool making) involve lateral asymmetries in the brains.

4.2.2.5. Implications of the Endocast Evidence

It appears that the processes of encephalization, lateralization, and reorganization that culminated in the modern human brain began with the earliest species of the genus *Homo,* and perhaps even somewhat earlier. Whether these changes were due to language, and therefore whether we can use them as a *terminus ante quem* for socially constructed coding, is not entirely clear.

It is risky for someone who is not a neuroscientist to try to evaluate such a question. Nevertheless, it appears to me that the work that has been done in this area falls short of proving the case. This is not because a good case cannot or has not been made for linking language to each of these traits, but because it has not been demonstrated that these traits are adaptations as opposed to exaptations for language. Humans are characterized by mental abilities other than the use of language. We have highly developed tool-making and tool-using abilities, for example, and we are capable of creating, manipulating, and rehearsing alternative scenarios in the process of planning or problem solving. It is not unreasonable, therefore, to ask whether the structures or morphological characteristics in question were the products of selection for one or more of these skills and were then co-opted by language.

Deacon, for example, makes a convincing case that the increased size of the prefrontal cortex is related to the ability to make and create linguistic symbols. Yet he devotes only a single paragraph (Deacon 1997:335-336) to the consideration of alternative explanations for the increase in prefrontal cortex, and he mentions neither tools nor scenario building. The role he cites for the prefrontal cortex is "maintaining attention on something in short-term memory in order to do something opposite or something complementary" (Deacon 1997:335). It would appear that this would be crucial for creating and comparing alternative scenarios in the course of planning. This is not to say that Deacon's hypothesis is not more probable than alternative hypotheses, but this is difficult to determine until it has been explicitly tested against them.

The same can be said for all the inferences made about language from endocast data. Jerison has not tested other possible uses for the increased computing capacity represented by encephalization in *Homo*: stone tool technology, planning, and even social demands other than language (Dunbar 1998; Reader and Laland 2002). Language functions are lateralized, but so are the motor operations involved in tool making and tool use.

The problem is particularly acute when macro-anatomical evidence for areas of the brain (such as Broca's area) are used to infer language. As Deacon put it (1997:287): "Though other primate brains have not evolved regions that are specifically used for language processes, those regions of the human brain that are did not arise de novo. The language areas are cortical regions that have been recruited for this new set of functions from among structures evolved for very different adaptations." Moreover, the equivalent of Broca's area has been found in nonhuman primates (Deacon 1997:342; Gibson 1996:409). Even if there is lateral asymmetry in such an area, it does not necessarily follow that this is attributable to language. Brodmann's area 44, which makes up a part of Broca's area, has a homolog in the great apes, and there is considerable left-right asymmetry in the size of this area in apes as well as humans (Cantalupo and Hopkins 2001).

From my perspective outside the neurosciences, therefore, it seems that the endocranial data, like the vocal tract data, are suggestive but not yet conclusive. Certainly the transition to a modern human brain began at least as far back as the first *Homo,* but until systematic and explicit testing of alternative hypotheses has been done, there can be no definitive demonstration of a link between endocast anatomy and language or socially created coding.

4.3. THE ARCHAEOLOGICAL EVIDENCE

The most striking and probably most reliable archaeological correlates of socially created coding consist of portable and parietal art, ritual burials, ritual objects and other evidence for ritual, and culturally determined style in artifacts. However, as I pointed out in the previous chapter, these are in fact products of the elaboration of culture. As such, they are also evidence of socially created coding, but unless we assume a priori that the earliest emergent coding appeared simultaneously with the elaboration of culture, then we should look for independent archaeological evidence for the earliest socially created coding. I do so in the remainder of this chapter. In the next chapter, I return to the evidence for cultural elaboration.

There are two places to look in the archaeological record for evidence for the origins of socially created coding. First, a number of authors have suggested that the manufacture and use of stone tools indicates, on various grounds, that their makers were capable of using, and probably were using, at least a simple form of language – language, of course, being a form of socially created coding. Second, it may be possible to find evidence in the archaeological record of the kinds of coordinated activities that would require socially constructed coding.

4.3.1. Stone Tool Technology as Evidence for Language

The archaeological record begins with the first surviving artifacts made by hominins. Although it has been argued that earlier objects of bone and antler were artifacts (Dart 1949; 1957; 1958; 1960; cf. Brain 1967; 1976; 1981), it is now generally accepted that the earliest indubitable hominin artifacts are the stone tools of the Oldowan industry, made by *Homo habilis* and *H. rudolfensis* and perhaps by one or more species of *Australopithecus* or *Paranthropus*. Three general classes of arguments have been advanced that stone tool making indicates the presence of language:

- The argument that language would be needed to teach stone tool making
- The argument that the making of stone tools involves the kind of arbitrary imposition of form that is characteristic of semantic symbolism and of grammar
- The argument that language and tool making require the same mental skills and that they therefore evolved together

By contrast, Noble and Davidson (1996) have asserted a very late origin for language based on the argument that the discovery of referential symbols underlies both language and technology involving significant planning (first evidenced by the colonization of Australia).

4.3.1.1. Language as a Requirement for Teaching Stone Tool Technology

Ingold (1994:285) cites Gowlett (1984a:55) as arguing "that the operations of early hominid toolmaking were of such complexity that they could only have been transmitted by means of language." This is actually a misquote; what Gowlett actually wrote was, "Language would have been of great assistance, especially for conveying cultural skills, but its possible presence and nature can only be postulated among the Oldowan tool-makers." Nevertheless, it raises the issue of how important language might have been to maintaining memetic traditions of stone tool making. There are two reasons to believe that Paleolithic knappers could have learned stone tool making perfectly well without language. First, Wynn (1993) has pointed out that even among modern humans, complex technological skills can be learned by observation combined with trial and error.[*] Second, the fact that modern professional and amateur archaeologists have repeatedly reconstructed prehistoric technologies (Johnson 1978) makes it clear that one can learn to make stone tools simply by looking at tools themselves, without even observing other knappers.

4.3.1.2. Stone Tool Technology and the Imposition of Arbitrary Form

Holloway (1969; 1973; 1976; 1981c) has argued that the making of stone tools involves an arbitrary imposition of form that is similar to the arbitrary imposition of meaning (i.e., symbolic meaning) involved in language and similar to the arbitrary conventions of what I call elaborated culture. In part, he bases this on the argument that if stone tools are standardized, then the standardization must be the product of cultural convention based on language. Thus he writes,

> The important question is whether or not other processes were also operating, such as *consensus,* or *explicit rules* about the forming processes. What is at issue is concatenated activity according to rules, i.e., grammar. Imitation and observational learning seem to me insufficient to explain the tremendous time depth and wide geographic extension of certain tool types in much of the Old World. It seems more likely that rules, consensus, syntax did exist, and that a communication system using symbolic language existed at least by the time of handaxes, if not before. (Holloway 1969:401)

---

* Gibson (1996) disagrees with Wynn on this point. However, she is arguing about the mental skills required to learn stone tool making, not about actual linguistic communication.

Isaac and Guilmet have made similar arguments about standardization. Isaac (1976a) linked increasing differentiation and standardization of stone tool industries through time to rule systems and "rising capacity for manipulating symbols, naming, and speaking." Guilmet (1977) argued that the use of language to pass on stone tool traditions would have resulted in more regular (standardized) tool forms.

Clearly, the foregoing quote indicates that Holloway and I have very similar notions of culture. What Holloway sees in standardized tool forms is essentially what I consider to be elaborated culture. For this reason, I will set aside until the next chapter the question of standardization, since all three authors are really using it as evidence of what I would call elaborated culture, not simply of socially created coding.

However, Holloway also argues that stone tool making involves the imposition of arbitrary form, because there is no resemblance between the raw material and the finished product:

> In the preparation of a stick for termite-eating, the relation between product and raw material is iconic. In the making of a stone tool, in contrast, there is no necessary relation between the form of the final product and the original material. (Holloway 1969:401)

The question here is whether socially created coding in the form of arbitrary convention can be inferred from the fact that the form of an unworked pebble does not suggest either the form of a flake struck from it or that of a core tool made by removing flakes from the pebble. Making flakes and flake tools depends on discovering that striking a flake from a core produces a sharp edge; making core tools depends on learning that removing several flakes from a core, especially from opposite sides of the same edge, also produces a sharp edge.

It seems to me that these are discoveries that can be made privately and learned memetically – by observation combined with trial and error. The relationship between pebble and artifact may be "non-iconic," but it is not arbitrary in the sense that a symbol is arbitrary. The kind of social interaction and consensus needed to create a commonly recognized symbol is entirely unnecessary for learning to create a flake. As Wynn (1991:197) expressed it in a slightly different context, "while a particular artisan may develop his own conventions and recipes, he does not share to any great extent with others. There are no rules to break, except perhaps his own or the limits imposed by the raw materials." Thus the making of stone tools does not, in itself, involve the imposition of form that is arbitrary in the sense of following socially created conventions (see Chase 1991; and Ingold 1994 for more detailed discussions).

4.3.1.3. Language and Stone Tool Making Require the Same Mental Abilities

A number of authors have argued that stone tool technology implies language because both require the same or similar cognitive skills or brain structures. Either implicit or explicit in this argument is the idea that the common psychological or neural basis of tool making and language indicates that the two evolved simultaneously. However, different scholars have come to this position from different directions and have cited different parallels between the two phenomena.

Leroi-Gourhan (1964a:164) and Montagu (1976) compared the procedure of tool making ("chaînes opératoires" in Leroi-Gourhan's terminology) to the grammar of language. Holloway and Kitahara-Frisch saw in the making of stone tools parallels to what Hockett called the design features of language (Hockett 1959; Hockett and Altmann 1968; Hockett and Ascher 1964). Kitahara-Frisch (1978) saw the making of tools for use at a later time as comparable to displacement in language (the ability to talk about things not present in one's immediate environment) and compared the use of tools to make other tools to reflexivity (the ability to use language to talk about language). Holloway (1969) found parallels in stone tool making to traditional transmission, duality of patterning (the fact that language is structured at both the phonological and the semantic and syntactic levels [see Chase 1991 for a critique]), and productivity (the ability of language to create new morphemes, words, and utterances).

A much more systematic and psychologically sophisticated argument was made by Gibson (1983; 1988; 1990; 1993; 1996; Parker and Gibson 1979), who explicitly linked the evolution of language and object manipulation on the basis of their relationship in human ontogeny. She used Piaget's (1952; 1954; 1955) framework of developmental stages in children, which "postulates that the maturation of intelligence proceeds through the *differentiation* of existing behavioral skills into finer component parts and the *combination* and recombination of separate skills into new and varied constructed, perceptual or behavioral wholes" (Gibson 1983:38, italics added). As a child matures, both object manipulation abilities and language progress through finer differentiation and more complex combination, with combinatory intelligence maturing later.

In addition, the neocortex is divided into different kinds of neural processing areas. Primary visual, auditory, motor, and somato-sensory areas provide differentiation within one "modality." Secondary areas provide synthesis or construction within a single modality, while association areas provide for cross-modal combination or construction. The primary and secondary areas mature together; association areas mature later. Moreover, "the highest constructional levels of both tool use and language are mediated by the same inferior parietal and anterior frontal association areas" (Gibson 1983:44).

Gibson presented three arguments for why the ontogeny of object manipulation and language could be taken as recapitulating their evolution. First, the skills developed at each stage are necessary building blocks for the next stage (see especially Parker and Gibson 1979). Second, the increasing size of the cortex throughout the course of hominid evolution made possible the increased number of neuronal connections that underlies enhanced constructional ability (see especially Gibson 1988). Third, the reorganization of the human brain relative to those of other primates involved not new structures but changes in the proportions (1) of the cortex to other parts of the brain and (2) among different parts of the cortex. Much of this proportional change is allometric in origin. The growth in size and elaboration of the cortex, which underlies human object manipulation abilities and language, is a late development in both ontogeny and primate evolution, and the association areas of the cortex mature latest (see especially Gibson 1996).

None of the arguments I have mentioned has gone unchallenged. While some design features of language seem to have analogs in tool making, others – notably duality of patterning – do not (Chase 1991). Wynn (1991; 1993) has thrown doubt on the parallel between grammar and the cognitive bases of tool making both on the basis of the way modern humans learn technologies and by pointing out that the creation of "rules" for making tools need not involve the kind of social interaction that creates symbolic language, including grammar.

The Piagetian model has been challenged both in general (e.g., Vygotsky 1962) and as it applies to the analysis of stone tools (Atran 1982). Moreover, Wynn (1989) applied Piaget's model to Oldowan tools without implicating language. Also, as Weaver et al. (2001) and Wynn (1991) have pointed out, the linkage between language and tool making is incompatible with theories of language as a specialized cognitive or neural module (e.g., Chomsky 1972; Mithen 1996a; 1996b; Pinker 1994).

Finally, it should be noted that positron emission tomography imaging of the brain of a modern archaeologist removing flakes from a core did not indicate that the portions of the brain usually associated with language were heavily involved (Stout et al. 2000). Because the test involved only one subject and one specific activity involved in making a tool, it cannot demonstrate that the evolution of language and tool making were not linked, but it certainly fails to confirm such a link.

It is difficult for an archaeologist without expertise in either psychology or neurology to evaluate the various positions taken in the literature. However, even if one accepts that language and tool making require the same set of mental skills or neurological traits not found among nonhuman primates, there are at least three possible ways to account for this fact:

1. Both may have evolved together.
2. Selection for tool making may have driven the evolution of abilities that later served as a preadaptation or exaptation for human language.
3. Selection for linguistic communication may have driven the evolution of abilities that later served as an exaptation for tool making (see Kitahara-Frisch 1978:108).*

It does not seem to me that any of the scholars who link stone tool making to language has made a methodical effort to consider alternative hypotheses by developing and applying test implications that would make it possible to choose among them. This does not mean that the first hypothesis is wrong, but because it has not been systematically tested against the alternatives, its validity remains in doubt. Stout, Toth, and Schick's PET imaging results reinforce this view.

4.3.1.4. Language Was Necessary for Technology Involving Significant Planning

Whereas the arguments already discussed place the origins of language very early in the archaeological record on the basis of stone tool technology, Davidson and Noble (1989; 1992; 1993; Noble and Davidson 1991; 1996; 2001) have concluded that language was a late development. As they see it, the first archaeological evidence for language is the colonization of Australia in the mid Upper Pleistocene (Davidson and Noble 1992; Noble and Davidson 1996:184), because this is the first evidence of the kind of depth of planning that would have required cognitive abilities that could not have existed without language.

Their argument is based on the idea that language provides "the capacity to adopt an attitude of reference to the perceived world" (Davidson and Noble 1989:125). This ability to reflect on what is perceived depends on the understanding that one thing can stand for another, as a linguistic symbol stands for a thing or concept. This understanding arose when iconic gestures were "frozen" through depiction, so that the icon could be understood to be a thing that stood for something else. Yet this understanding had to come into being in the social context of lan-

---

* It is also logically possible that the same underlying abilities evolved separately in response to selection for language and for tool making. However, the fact that the same association areas of the cortex are involved in both makes this unlikely.

guage, so that reflective thinking not only depends on language but is in a way a part of language.

Thus they propose that "against the view … that 'consciousness,' 'thinking,' 'imagination,' 'memory,' and so forth, are categories of the mental, independent of language, we propose instead that these terms are grammatical categories" (Davidson and Noble 1989:132). It is only the ability to objectify referential signs (symbols and icons) that makes it possible to create, rehearse, and manipulate the kinds of mental scenarios involved in planning. They draw on the philosophical and psychological writings of Wittgenstein, Ryle, Coulter, James Gibson, Vygotsky, and others when they argue that that mental representations are impossible without language because the "mind" does not exist. Rather, "mindedness" is a product of shared meaning and therefore of language.

Although the archaeological record contains direct evidence for symbolism in the form of art, ritual, and so forth, the earliest solid evidence is the colonization of Australia. The use of boats is inferable because even at low sea levels a 90 km stretch of open water had to be crossed.

> Balme (1990) has argued, plausibly, that humans in the islands to the north-west of Australia were using boats for fishing before they used them to get to an unknown Australia. There are at least three layers of intentionality here: the act of making the boat involves what Mellars (1989) called 'imposed form', with the fashioning of raw materials in ways not entirely determined by the mechanics of manufacture; the act of combining different kinds of raw materials (e.g., fibre and bamboo stems) into a single artifact; and the act of making the boat for the purpose, which, at the time of manufacture, could only have been imagined. These acts involve what we might call 'mental representation' beyond that required by the other products of behaviour which can be observed or inferred from the same or earlier periods. (Davidson and Noble 1992:138)

Thus the colonization of Australia is a *terminus ante quem* for the origins of language (and, for our purposes, of socially created coding). Noble and Davidson (1996:217) conclude that "sometime between 100,000 and 70,000 years before the present the behaviour emerged which has become identified as linguistic."

I am personally skeptical of their position, but this is not the place for a systematic evaluation of their philosophical arguments. Many scholars who are better qualified than I hold views in direct opposition to Noble and Davidson's (e.g., Chomsky 1972; Churchland and Sejnowski 1992; Dennett 1991; Fodor 1975; Fodor 1983; Gallistel 1989; Pinker 1994; Spitzer 1999; see Carruthers 2002 for a summary of philosophical positions on the relationship between language and mind;). This does not mean that Noble and Davidson are wrong, of course. It does mean that archaeologists cannot accept their arguments uncritically.

Although Noble and Davidson believe that mental representations are impossible without language, there seems to be good evidence that nonlinguistic species are capable of creating mental representations.

> The other defining characteristic of cognitive processes is some form of mental *representation:* In its decision making process, the individual relies on information from a source other than direct perception, for example, from memory, inference, categorization, or insight. Thus, when an animal is faced with a problem and does not engage in overt behavioral trial-and-error but solves the problem "intelligently" nonetheless, in many cases the basis of that intelligence is some form of mental assessment of the situation. The organism compares the current situation to a previously experienced situation or infers what will happen if certain aspects of the situation are changed. In these cases the individual seems to make an assessment by manipulating the elements of some internal mental model of the problem (i.e., its knowledge) and observing the imagined consequences in that model, as, for instance, when a rat decides on the direction in which it should forage based on an assessment of the likelihood of finding food in the different directions and the possible obstacles to its path that each direction presents. When representation is combined with behavioral flexibility in a dynamic form of mental assessment, we get the prototype of cognitive processes called "thinking." There is no reason to doubt that many animals engage in thinking when it is defined in this way [Tolman, 1948]. (Tomasello and Call 1997:10)

Tomasello and Call go on to provide a good summary of the evidence, based on both spatial decision making and tool use or tool making, that primates other than humans are using mental representations in this way. In fact, Noble and Davidson (1996:128) acknowledge that chimpanzees "seem able to acquire 'some form of internal representation'."

In the current context, however, it does not really matter whether Noble and Davidson are right or wrong. The question at hand is not whether hominins were using language to communicate and to think by the time people reached Australia. If Noble and Davidson are right, then the peopling of Australia marks a *terminus ante quem* for language. Thus, even if planning on the level required for the peopling of Australia was impossible without language, it does not follow that socially constructed coding did not date back to the Middle Pleistocene. To draw such a conclusion, it would be necessary to demonstrate that language could not exist without being visible in the archaeological record or, at the very least, to show that claims for language based on skeletal anatomy were not just inconclusive but wrong.

4.3.1.5. Is Stone Tool Making an Indicator of Socially Created Coding?

It appears to me that at this point the case for inferring language from the presence of stone tools in the archaeological record is not proven. As should be clear from the foregoing discussion, I have three different grounds for this conclusion.[*]

First, although there are resemblances between language and stone tool making, none of these necessarily involves socially created coding. I think claims that stone tool making per se involves either imposition of arbitrary form or duality of patterning are mistaken.

Second, even when it is clear that specific cognitive abilities or neural structures are common to both language and stone tool making, the possibility that these evolved in the context of tool making and became an exaptation for language seems not to have been systematically tested and disproven.

Third, when specialists disagree on topics on which I as an archaeologist am unqualified to pass judgment, I am loathe to accept any one argument as demonstrated. This may seem like vacillation, but the relationship between language and stone tool technology is fundamentally a psychological or neurological, not an archaeological, problem. It will have to be settled by specialists within those disciplines.

4.3.2. Archaeological Evidence of Coordinated Activities

We know that in recent times, many of the economic activities of hunter-gatherers have been cooperative endeavors coordinated by means of socially created codes – cooperative hunting, foraging, or food processing, food sharing according to socially constructed rules, and so forth. Since of all hominin behavior, economic activities are probably the most likely to leave traces in the Paleolithic archaeological record, it should in theory be possible to find evidence of such coordination. While the search for such evidence reminds us how little we really know about the

---

[*] I have omitted a few studies from the preceding discussion because their applicability to the question at hand is uncertain. Ingold (1994) distinguished between technique and technology and argued that the latter was possible only with language. Although his idea is of considerable interest, it is not clear how the distinction would be reflected in the archaeological record. Donald (1991) argued that the adaptation of *Homo erectus* was characterized by mimetic (not to be confused with memetic) communication and culture. He was not, naturally, concerned with emergent coding when he wrote this book, so it is unclear whether the kind of mimesis he attributed to *Homo erectus* would necessarily involve or imply emergent coding. Hewes (1973) wrote about the relationship between tool making and language, but the main thrust of the article was that a gestural origin of language would make it easier to establish a link between the two than would a vocal origin. He did not present a detailed argument about why language and tool making must have evolved together.

details of Paleolithic behavior, there are nonetheless some reasonably believable clues. There are also some less reliable clues.

4.3.2.1. Compound Tools and Social Interaction

Reynolds (1993) argued that two characteristics were unique to human tool technology as opposed to that of nonhuman primates. First, humans create compound tools by attaching two or more separate components so that they become part of a single whole – for example, by hafting a point on a shaft to make a stone-tipped spear or javelin. (Nonhuman primates may pile one object on another, but they do not attach objects to one another without human assistance.) Second, among humans, tool making and tool use is, habitually, a social rather than a solitary activity, with the actions of one person complementing and often anticipating those of another.

There is little reason to disagree with either of these assertions. Of course, Reynolds was not concerned with socially created coding per se and did distinguish coding and behavior, but his ethnographic descriptions of tool making and tool use among modern humans paint a compelling picture of the social nature of human technology. For example, he described a videotape of two Australian men making stone knives with handles molded from grass seed resin:

> I was struck by the complementarity in the actions of the two men. With little fanfare, each individual would anticipate what the other was about to do and facilitate it by performing the complementary action. In the preparation of the thermoplastic to be used as the handles for the stone knives, one of the men emptied the spinifex seed from a modern plastic bag into a traditional wooden dish, about a meter long and shaped like an elongated oval, and then smoothed out the seed to an even depth. Although one man already was gripping the tray at opposite ends with both hands, the other grasped the tray as well, and the two of them together lowered it onto the fire to heat. Also, in tending the fire, one man handed a stick to the other, who in turn raked the coals. ... Then as the fire began to wane, one of them lifted the dish so the other could stir up the blaze underneath by putting more fuel on the fire. The next scene begins with cooperative stirring of the thermoplastic which has formed in the dish. One of the men takes the warm, viscous liquid and molds it into a ball about the size and shape of a baking potato. ... Then the two men divide the labor into two tasks, with one of them molding the thermoplastic into roughly shaped knife handles, while the other gives them their final form and attaches them to the stone knife blades that had been prepared earlier. (Reynolds 1993:411)

It seems certain that these two men were acting according to commonly held codes and that their behavior was social in nature. It could be argued, however, that such social *behavior* is possible on the basis of me-metic rather than socially created coding.

Reynolds goes on to say that these two traits of modern human tech-nology "are co-evolutionary developments, emerging together in syner-gistic fashion" (1993:422), and states that this can be tested archaeologi-cally: "one would expect arrangements of objects supported by gravity to co-occur with simian fossils, since they occur in contemporary apes, whereas the presence of polyliths [true compound tools] would be defini-tive of *Homo*" (Reynolds 1993:424).

This co-evolution seems to me unproven. The test is not whether compound tools are found only with *Homo,* since this proves only that other fossil primates did not create compound tools. (After all, the first stone tools in the archaeological record were made exclusively by hominins, yet they are not compound tools, since there is no evidence that they were hafted.) Neither an archaeological association of com-pound tools with *Homo* nor examples of people making compound tools in a social context today are in any way proof that the earliest creation of compound tools required or involved socially created coding rather than memetic coding – or even that it involved social interaction at all. In other words, just because technology is a social activity today, it does not follow that compound tool technology either requires or originated in conjunction with socially created coding.

4.3.2.2. Artifact Concentrations, Home Bases, and Cooperation

Isaac (1976b; 1978) noted that the Oldowan archaeological record consisted of concentrations of both animal bones and stone tools and debitage. On the basis of ethnographic analogy, he interpreted these as home bases, to which early *Homo* transported animal carcasses or parts of them, vegetal foods, and the stones with which to process them. Be-cause females encumbered with young would have been handicapped relative to males in terms of hunting or of scavenging in competition with large carnivores, Isaac inferred a sexual division of labor. Meat and vegetal foods would have been brought back to the base camp by mem-bers of the two sexes and shared there. This pattern of reciprocal, com-plementary economic activity would have been a major change from the individualistic foraging of the apes. From our perspective, this would have required both cooperation and coordination of activities.

Potts (1984; 1987) pointed out that the internal structure of Oldowan stone and bone concentrations differed in significant ways from the structure of ethnographically known home bases. In the Oldowan depos-its, bones had evidently accumulated over years, with continuing carni-vore involvement. Moreover, the bones had not been as intensively util-ized as in ethnographically known home bases. All this implied not a central base from which hominids foraged and to which they returned to share food, but locations visited sporadically over many years. Each visit brought a very limited amount of meat and bone, which was not heavily utilized.

Setting aside natural accumulations, several alternative models could account for such a pattern: hominins' bringing both bones and stones to some more or less permanent location, such as a shade tree, that provided shelter, safety, or some other amenity; their use of a carnivore bone accumulation; their scavenging at locations where animals frequently died, either from carnivore hunting or other causes (Isaac 1983); or their bringing parts of carcasses for quick processing to stone caches established about the landscape as a way to avoid competition from carnivores (Potts 1984; 1987).

Although these considerations make it impossible to be sure that central place foraging was a characteristic of very early *Homo,* by the Middle and Late Pleistocene there were certainly sites that resemble the home bases of ethnographically known hunter-gatherers. Such sites imply frequent transportation of resources, probably fairly intense occupation (at least on the order of days rather than hours per visit), and quite likely food sharing as well. However, other species also bring food to their offspring at a fixed home base. It is likely, but probably impossible to demonstrate, that such sites indicate sharing of food rather than simply provisioning of the young.

Even if sharing was extensive among early hominins, it is difficult to know whether this pattern of central place foraging and food sharing required socially created coding to coordinate it. The very fact that a base camp exists makes it easy to coordinate foraging, because everyone knows where to go at the end of the day. Such behavior is common among species such as social carnivores, who return to their dens or to the location where the young are located, without benefit of socially created coding. Yet even the coordination of movement to locations other than the base camp can apparently be accomplished without socially created coding.

The Hamadryas baboons of Erer-Gota, Ethiopia, have a daily routine according to which in the morning they leave a sleeping cliff in any one of many directions. They travel and forage until late morning, when they stop at one of a number of waterholes surrounding the cliff, and they return to the same sleeping cliff by evening (Sigg and Stolba 1981). These baboons apparently coordinated their movements during the day, even when the band had divided into separate groups.

> The different parts [of the band] were out of sight of each other, but met again after some time. In 5 cases, the separate parties were pursued by 2 observers; the parties followed quite different routes, e.g.,

one part crossed a hill chain, while the other part went around it. Nevertheless, they later met at the same rich feeding site, despite other feeding sites situated closer to the route of one party. The particular conditions made it unlikely that the baboons met again through optical or acoustic communication; rather, the events support the hypothesis that the baboons know in advance where to go, and where others will go. (Sigg and Stolba 1981:71)

Thus, like so much of the other evidence already considered in this chapter, the archaeological evidence for apparent home bases in the late Middle or Upper Pleistocene is suggestive but not conclusive of socially created coding.

4.3.2.3. Cooperative Hunting and Socially Created Coding

While much of Paleolithic economic life remains invisible because little vegetal material survives in the archaeological record, the bones and teeth of prehistoric prey do give us an idea of hunting activities. Among recent humans, hunting is often a cooperative activity, and in many cases it involves extensive planning and coordination that would be impossible without socially created coding. We know little about the details of how Paleolithic humans actually killed game before the late Upper Pleistocene. A few wooden spears are preserved (Oakley et al. 1977; Thieme 1997), one of them from the rib cage of an elephant (*Palaeoxodon [Elephas] antiquus*) from Lehringen in Germany that was apparently killed and butchered by humans (Adam 1951; Movius 1950; Thieme and Veil 1985; Weber 2000). It has been argued on the basis of use wear that at least some Middle Paleolithic points were used as projectile points (Shea 1988; 1997), and a Levallois point was found embedded in the vertebra of an ass (*Equus africanus*) at Um el Tlel, Syria (Boëda et al. 1999).

Nevertheless, even the killing of large game such as an elephant is not in itself evidence of coordination requiring socially created coding, because we do not know the circumstances under which the animals were killed. Elephants might have been scavenged rather than hunted in many circumstances. Even when, as in the case of Lehringen, the animal appears to have been killed, it is possible that it was severely weakened and vulnerable. (Another *Palaeoxodon antiquus* at Gröbern, Germany, that was apparently butchered by humans was in poor health when it died [Mania 1990:221]).

It is possible that ambush hunting or other methods could have been used on rather large and dangerous game without a great deal of coordination among the hunters. The same is not true of driving herds of animals over jumps or into enclosures.

Lions and wolves may surround herds in order to attack individuals within them, but they do not attempt to move herds or groups of animals as units, guiding them to a place where they can be killed off en masse. This appears to be something done only by humans. Ethnographic descriptions of large game drives indicate that they involve a high degree of coordination among the hunters (e.g., Arthur 1975; Blehr 1990; Gordon 1990; Nellemann 1969/70; Verbicky-Todd 1984). Indeed, it is difficult to conceive of how such a drive might be conducted without participants' first coming to at least some basic agreements about the direction in which the game were to be driven and who was to do what to accomplish this.

Because such agreements would constitute socially agreed-upon coding, and because they would likely involve language, it seems to me that archaeological evidence for driving herds also constitutes fairly strong evidence for socially constructed coding.

When herds of animals are driven into natural or artificial traps and killed as a herd, the age profile of the dead animals naturally reflects that of the living animals, with young animals more common than prime age adults, which in turn are more common than older animals. Thus an age profile dominated by young animals, followed in frequency by prime age adults, with older animals increasingly scarce (Klein et al. 1983; Lyman 1987) should in theory be a good indication of hunting by driving. Other forms of hunting are likely to be selective in terms of the ages of the animals exploited. Even when the weakest animals are killed, these are usually the youngest and oldest individuals in the population.

In practice, the age profile produced is slightly different, for taphonomic or other reasons. In Holocene bison kills from North America, where the circumstances and features of the sites and their vicinities make it clear that herds of animals were driven to their deaths, very young animals are underrepresented (Speth 1997).

A number of early Upper Pleistocene sites have yielded age profiles of large herd animals that match just such a catastrophic profile (with the very young underrepresented):

- Age profiles of eland (*Taurotragus oryx*) from the Klasies River Mouth and Die Kelders Cave 1 sites in South Africa matched a catastrophic-death age profile, indicating that they were killed by driving (Klein 1979; 1987; Klein and Cruz-Uribe 1996). Both sites date to the early part of the Upper Pleistocene, Klasies River Mouth to oxygen isotope stage 5 (128,000–74,000 years ago) and stage 4 (74,000–59,000 years ago), and Die Kelders 1 to stage 4 (Klein and Cruz-Uribe 1996:318).
- In France, the site of Mauran yielded the remains of between 98 and 137 individual bison (*Bison priscus*) from only 25 m^2 exca-

vated in a site estimated to cover 1,000 m^2. The ages of these animals fit a catastrophic age profile, indicating driving (David and Fosse 1999; Farizy et al. 1994). However, this was not a mass drive. Rather, microstratigraphic analysis indicated repeated kills of smaller numbers of animals (Farizy et al. 1994:239). Mauran probably dates to the last interglacial, stage 5b or 5c (Farizy et al. 1994:69-70).

- The site of La Borde in France yielded a catastrophic mortality age profile for aurochs (*Bos primigenius*) found with a Denticulate Mousterian industry (Jaubert et al. 1990). The site dates to a warm stage, either the last interglacial (stage 5) or an earlier interstadial.

- Age profiles of horse (*Equus*) teeth from beds 14, 22, and 23 of Combe Grenal, France, also indicated killing by driving (Levine 1983). These levels dated to oxygen isotope stage 3 (Laville 1988; Mellars 1988) and were associated with Denticulate (bed 14) and Quina (beds 22 and 23) Mousterian industries.

- Il'skaya, in the northern Caucasus, produced a catastrophic age profile for bison, based on the upper first molar (M1), or a prime-age-dominated profile, based on the lower third molar (M3) (Hoffecker et al. 1991). The latter is essentially the same as a catastrophic profile with the young underrepresented. Since M3 erupts very late, however, M1 probably provides a more accurate profile. As at Mauran, the animals were killed in multiple drives over a period of time. The site was dated to stage 5d–5e. (It should be noted that the Il'skaya assemblage was the product of excavations in the 1920s and 1930s.)

An even more striking set of data comes from the site of La Cotte de Saint Brelade in Jersey. There, two separate levels yielded evidence of mammoth (*Mammuthus primigenius*) drives (Scott 1980). The remains were in a narrow gap or gorge in the rock face of what, at the time of occupation, was a headland overlooking the coastal plain. The location of the remains made it virtually certain that they were not brought to the site by carnivores. The height of the site above sea level and the nature of its fine loess sediment eliminated both wave and stream action as the accumulating agent. Finally, although the condition of the bones indicated that they were buried rapidly, they were not uniformly represented. Rather, different parts of the animals appear to have been moved to different parts of the site. In level 6, scapulae were stacked up in one area. Bones were piled leaning against each other.

The number of individual animals was small from a statistical point of view, but both teeth and bones produced age profiles consistent with humans having driven groups of animals over the cliff. There was no evidence of intensive occupation of the site, which might have implied that this was a home base to which hunters carried parts of carcasses (after all, who would want to drag a mammoth skull up a headland?). Rather, the small number of artifacts found was more consistent with butchering at a kill site. Thus all the evidence points to repeated drives of groups of mammoths over the cliff from the plateau above La Cotte de Saint Brelade. The site dates to a cold period before the last interglacial – in other words, to the Middle Pleistocene (McBurney and Callow 1971; Scott 1980).

Thus it appears that by the late Middle Pleistocene, hunting of large game by drives was an established behavior. It seems unlikely that such drives could have been organized and carried out without some form of socially created coding to coordinate the actions of those involved. Although it is always risky for archaeologists brought up in an industrial economy to guess just what might or might not have worked for Pleistocene hunters, it is hard to imagine how this kind of coordination could have been achieved without some form of socially constructed coding, probably including language.

4.3.2.4. Mortuary Evidence

Hominin crania from two Middle Pleistocene sites in Ethiopia, Bodo and Herto, bear apparent traces of defleshing (Clark et al. 2003; White 1986; White et al. 2003). According to White, these are unlikely to have been the results of nutrition-oriented cannibalism. If he is right, then these specimens imply the occurrence of some kind of symbolic mortuary practices, which in turn would support the conclusion that the social creation of coding dates at least to the Middle Pleistocene. (Why it does not also imply the elaboration of culture into all-encompassing systems will be addressed in the next chapter.)

4.4. Conclusion

It has proved frustratingly difficult to pinpoint the origins of socially created coding on the basis of research done to date. To a considerable extent, this is due to weaknesses in the chain of arguments linking data to hypotheses. In some cases, researchers have concentrated on providing supporting arguments for their own hypotheses and have made no systematic attempt to evaluate alternative hypotheses. More often, crucial links in the chain of argument are simply unavailable. No one really understands, for example, the causal link between basicranial flexion, hyoid position, and the use of formant frequencies in modern human speech.

The problem is seriously exacerbated, of course, by the fact that we cannot observe Pliocene and Pleistocene behavior. We have no Pliocene or Pleistocene brains to study, and of all the materials used or modified by Paleolithic hominids, only stone and bone have survived in any quantity. Strictly speaking, therefore, we simply cannot establish the earliest appearance of socially created coding, or even the date at which socially created coding became a significant part of hominid adaptation.

However, all is not lost. For one thing, the significant number of zooarchaeological assemblages with age profiles indicative of the driving of large game seems to me to provide a *terminus ante quem* for socially created coding. If I am right, then we have archaeological evidence of socially created coding by the late Middle Pleistocene.

This is compatible with other lines of evidence discussed earlier. Although the validity of none of these has been demonstrated, taken together they are suggestive. By the end of the Middle Pleistocene, the hominin brain was of modern size and, macroscopically at least, of modern organization, with essentially modern patterns of lateralization and with modern language-related landmarks. By the Middle Pleistocene, at least some hominins had modern basicranial flexion. Because the evolution of adaptations for speech must have lagged the origins of language by a considerable time, the absence of modern basicranial flexion in Neanderthals does not indicate an absence of language in that population. By the Middle Pleistocene, the archaeological record is replete with sites whose composition is compatible with base camps equivalent to those known among modern hunter-gatherers – and this may imply coordination of behavior along the lines of modern hunter-gatherers. The origins of language and socially created coding may lie much farther back than the late Middle Pleistocene. The size increase, reorganization, and lateralization of the brain began much earlier, and the use of formant frequencies as an adaptation must have lagged the evolution of language.

Thus, it appears to me that the evidence suggests that socially created coding was playing a significant role in hominin adaptation by the late Middle Pleistocene, and perhaps much earlier. I am not particularly happy with the solidity of the arguments and data underlying this conclusion, but it seems that, for the moment, it is the most justifiable one to reach.

5

THE ELABORATION OF CULTURE

In the previous chapter, I reviewed the evidence regarding the origins of socially created coding, including the origins of referential language. In this chapter, I review the archaeological evidence for the elaboration of such coding into overarching, all-encompassing systems of symbolic culture. I use the archaeological data to evaluate the three alternative hypotheses set forth in chapter 3 as possible explanations for this elaboration of culture. These hypotheses were the following:

- The "by-product" hypothesis, that the elaboration of culture was an adaptively neutral spandrel or by-product of the evolution of socially constructed coding
- The "anxiety" hypothesis, that the elaboration of culture can be explained as a mechanism for allaying individual emotional anxieties
- The "group benefit" hypothesis, that the elaboration of culture is a means of motivating altruistic behavior that benefits the group, even at the expense of the individual.

In discussing how to test these alternative hypotheses against one another,

- I argued that if either the by-product hypothesis or the anxiety hypothesis was true, then it followed that culture must have been fully elaborated as soon as socially created coding became a part of the behavioral repertoire of hominins. The only hypothesis that allows for any significant time gap between the origins of socially created coding and its elaboration into all-encompassing systems is the group benefit hypothesis.
- I also argued that, since the use of simple referential symbolism would be for the most part archaeologically invisible, most ar-

chaeological evidence for symbolism was actually evidence for the elaboration of culture.

Thus the question to be addressed in this chapter is whether or not the elaboration of culture occurred essentially simultaneously with the origins of socially created coding.

My tentative conclusions about the timing of the latter phenomenon provide a working premise for this chapter. In chapter 4, I reviewed the osteological and archaeological data bearing on the origins of socially created coding. Although the evidence is by no means incontrovertible, I accepted that it does suggest that socially created coding was part of the hominin adaptation by the end of the Middle Pleistocene. This will serve as a basis for testing the three hypotheses I have proposed for explaining the elaboration of culture into overarching, all-encompassing symbolic systems. If the elaboration of culture dates back to the Middle Pleistocene or earliest Upper Pleistocene, then the group benefit hypothesis will have to be rejected. If there is a time gap, however, the other two hypotheses will have to be rejected.

The archaeological evidence that I review here has been the center of an extensive and sometimes fierce debate about the origins of symbolism (e.g., Bar-Yosef 1988; Bednarik 1992; Belfer-Cohen 1989; Chase and Dibble 1987; 1992; Davidson 1991; Davidson and Noble 1989; Duff et al. 1992; Facchini 1998; Hayden 1993; Kaufman 2002; Klein 1999; 2000; Knight et al. 1995; Lindly and Clark 1990; Marshack 1985; 1988; 1990; Mellars 1973; 1996a; 1996b; Noble and Davidson 1991; 1996; Stringer and Gamble 1993:203-207; Watts 2002; White 1982). The evidence includes early objects claimed be artistic or symbolic in nature, objects or configurations of objects claimed to represent evidence of ritual or religion, and patterning of artifact morphology claimed to represent stylistic variability.

This debate has aroused strong feelings. Many archaeologists and human paleontologists have a stake in it, and in many cases the positions of the protagonists are fixed and even hardened. I must therefore emphasize very strongly that the issues involved in the debate over symbolism and the questions that concern me in this chapter are very different. In this chapter, I am *not* doing any of the following:

- Investigating the origins of symbolism
- Investigating the origins of "modern behavior"
- Investigating the relative intelligence or cognitive capacities of anatomically modern humans and Neanderthals (or any other group of hominins)
- Investigating the origins of an aesthetic sense

- Investigating the origins of an ability to recognize or to appreciate iconic resemblances[*]

All I am doing is testing the three hypotheses proposed in chapter 3 to explain the elaboration of culture into overarching and all-encompassing systems. The data must consequently be regarded from a very different perspective than that needed for the investigations just listed. It may be that my discussion of the data has some relevance to the goals just listed. If so, that is fine. But I ask readers to set aside their usual perspectives on the material discussed here and to consider it in the context of testing these three hypotheses.

To begin with, if one accepts the arguments I made in chapter 3, then the implications of these kinds of data for the evolution of hominin intelligence are nil. Most archaeologists (including myself prior to 1999) have made no distinction between referential symbolism and elaborated, all-encompassing symbolic cultural systems. As a result, we have assumed that the archaeological evidence for symbolism tracks the use of symbols (including the earliest referential symbols) rather than the elaboration of culture.

The use of symbols implies a certain level of cognitive or neural evolution, in terms either of general intelligence or of specific neural structures essential to symbolism and language. This linkage between symbolism and intelligence is probably the reason the debate about this kind of archaeological evidence has become so impassioned. Many archaeologists feel that questioning such evidence in the archaeological record of any given population (such as Neanderthals) can be seen as demeaning that population and questioning its humanity, because questioning the data amounts to questioning intelligence. (For especially explicit expressions of this perspective, see Belfer-Cohen and Hovers 1992:470; Corruccini 1994; Hayden 1993).

Such a stance, however, requires three assumptions: (1) that symbolism requires an evolved intelligence, (2) that the archaeological evidence marks the first use of symbolism, and (3) that any hominins with sufficient intelligence to use symbols will do so. While the first assumption is probably justifiable (although we must remember that other living primates apparently have at least some symbolic abilities), the third is open to question, and I argued in chapter 3 that the second is not a valid assumption. If this is right, then arguing that the archaeological record left by any given hominin population contains no evidence of elaborated cul-

---

[*] I use the term "iconic" in Peirce's (1932 [1960]) sense as a thing that indicates or points to something else by resembling it. Whatever psychological mechanisms may be required for recognizing such resemblances, icons do not depend for their meaning on the socially constructed conventions involved in symbolism.

ture says absolutely nothing about those hominins' intelligence, about their capacity for symbolism, or even about whether or not they were using referential symbols as an inherent part of their adaptation.

Certain behaviors have been identified as "modern" by archaeologists (especially by Africanists, ironically) primarily because they are characteristic of the European Upper Paleolithic – for example, blade technology, bone and antler working, compound tools, efficient hunting of large game, fowling, and fishing (Deacon 2001; Klein 1999; 2000; McBrearty and Brooks 2000; see also Henshilwood et al. 2002). Yet none of these has been linked other than by association with the European Upper Paleolithic to elaborated systems of culture, and there is nothing inherent in them that would require such a link. Until someone can demonstrate why such a link is necessary, archaeological evidence for the presence or absence of such behavior at any given place or time is irrelevant to the question that concerns us here.

5.1. CRITERIA FOR EVALUATING THE EVIDENCE

Discussion of the archaeological material considered in this chapter has always been lively, and at times remarkably acrimonious. One thing has, unfortunately, become clear to me. Not only is there no agreement about the evidence itself, but there is no agreement about the criteria by which the evidence is to be evaluated. Very likely, there will be no agreement on either point during my lifetime, for several reasons.

First, a number of very different epistemological traditions are involved in the debate. Most Paleolithic archaeologists from English-speaking countries have at least a rough and ready scientific approach to the data, in the sense that they try to work in terms of testable hypotheses for explaining the archaeological record. (To date, postmodernist voices have played only a minor role in the debate, at least within archaeology.) Many Continental European archaeologists come from quite different epistemological traditions, including the inductive natural history approach against which North American archaeologists rebelled explicitly in the 1970s. Others follow a rationalist approach, in which data are not used to test hypotheses but are pegged into logically constructed a priori frameworks. In addition, scholars from other fields such as philosophy, linguistics, and art history have also been involved, and they often have different epistemological perspectives from those of archaeologists.

Second, interpreting the data often depends on making probabilistic judgments when adequate information is simply unavailable. For example, the argument has often been made that Middle Paleolithic people were burying their dead in caves and rockshelters, because so many intact or largely intact skeletons and/or skulls have been recovered in such sites. As Gargett (1999) pointed out, however, it is difficult to get reliable comparative data to establish whether such a probabilistic argument is valid. The frequency of preservation seems to be high relative to that of non-hominin remains found in caves and shelters. Non-hominin carcasses and skeletons represent the debris of human butchering and consumption or else have been heavily damaged or disturbed by carnivores (or both). It cannot be ruled out that dead hominins, unlike dead ungulates, were left unbutchered in caves and shelters from which carnivores were in large measure excluded by the repeated human occupation.

The fact is that we do not know whether carnivores avoided caves habitually occupied by humans even when the humans were not actually there, and we do not know whether cadavers left on the surface in a site habitually reoccupied by humans would have been significantly less damaged and disturbed by subsequent occupations than the carcasses of game animals. My own opinion, colored by many hours spent studying faunal remains from both heavily and sporadically occupied sites, is that the kind of preservation found in hominin remains is unusual. I would be the first to admit, however, that this is a subjective judgment, and that I do not have the reliable quantitative data that would be needed to defend it objectively.

It is inevitable, therefore, that my review of this kind of evidence will hardly represent the last word on the subject. Indeed, I hope it will not. I would far prefer that it be the first of a series of debates about the elaboration of culture.

With this in mind, I review the currently available evidence from the Middle and early Upper Pleistocene. First, however, I explain the criteria I use to evaluate the data and which classes of data I do or do not consider relevant in light of these criteria.

5.1.1. The Principle of Simplicity

In the following discussion, I adhere to one principle that I expounded with Dibble in the context of the old debate about the origins of symbolism (Chase and Dibble 1992). When two interpretations of a datum or set of data are possible, I prefer the simpler one. There are two sides to what constitutes a simpler explanation. On the one hand, an explanation is simpler if it invokes only factors present in nature, among all primates or among all hominins. Explaining something in terms of either natural, taphonomic factors or in terms of practical functions that benefit the individual is simpler than explaining it by invoking socially created coding and the elaboration of such coding into all-encompassing cultural systems, because there is no resort to evolutionarily new behavior. Dibble and I made the same point with regard to early hominin behavior:

> A good archaeological example of a rich versus a simple explanation
> is provided by Dart's (e.g., 1957) hypothesized "osteodontokeratic"
> industry, a supposed pre-lithic tool kit based on skeletal material.
> Dart used this concept to account for certain discrepancies in the rela-
> tive frequencies of skeletal elements and for certain patterns of bone
> breakage in South African australopithecine-bearing deposits. Such
> an explanation was very satisfying to many who believed (1) that tool
> use was an important impetus to early hominization, and (2) that a
> phase based on naturally shaped objects was logically prior to the de-
> liberate manufacture of stone tools. However, it was later discovered
> that both the biases in skeletal composition and the patterns of break-
> age that he cited could also be produced by natural taphonomic fac-
> tors, especially carnivore accumulation and destruction (e.g. Brain
> 1969, 1981). This taphonomic explanation is the simpler one, since it
> depends only on natural factors already known to operate on bones,
> and does not depend on a new behavior on the part of the hominids. It
> is interesting that most archaeologists have accepted it as the better
> explanation, even though no-one has ever proven that the skeletal
> material in question was *not* used by the hominids. Following the
> same logic, then, a taphonomic explanation of an archaeological phe-
> nomenon is simpler than one that invokes symbolic behavior. (Chase
> and Dibble 1992:45-46)

On the other hand, a taphonomic or functional explanation may be, internally, extremely complex. A straightforward explanation that evokes the
behavioral and social complexity of fully elaborated culture may thus be
simpler than a convoluted and improbable taphonomic or functional one.
Determining which explanation is simpler is a judgment call, a matter of
balancing two opposing considerations.

5.1.2. The Problem of Taphonomy

The kinds of archaeological materials cited as early evidence for
symbolism – or elaboration of culture – are often ambiguous. In such
cases, there may well be honest disagreement about whether the item in
question is a product of human workmanship or of taphonomic processes. One good example comes from the site of Divje Babe I in Slovenia. A partial femur of a cave bear was recovered from a layer dated to
about 43,000 years ago (Turk 1997a). It has two holes in one surface,
aligned along the axis of the bone, and a deep notch, part of which could
represent a third hole. It resembles a flute, but even after a great deal of
investigation and debate, archaeologists have unable to agree on whether
or not the holes were the product of carnivore gnawing or of deliberate
human manufacture (Albrecht et al. 1998; Albrecht et al. 2001; Bastiani
1997; Chase and Nowell 1998; D'Errico, Villa et al. 1998; Horusitzky
2003; Nowell and Chase 2002; Turk et al. 2003; Turk et al. 2001; Turk et
al. 1997a; 1997b). Perhaps more significantly, most of those involved in

the discussion agree that it is impossible to exclude either possibility with certainty.

Other objects are less ambiguous. Most notable is the abundance of punctured artiodactyl phalanges that have sometimes been interpreted as whistles. Good evidence exists that many of these were created naturally, usually by carnivore gnawing, and I know of no solid evidence that any of them was made by humans (Chase 1990; 2001c; Lartet and Christy 1865-1875:44, B pl. V; Martin 1906; 1909:150-68; Stepanchuk 1993; Wetzel and Bosinski 1969:126-8). Bear teeth with grooves in the dentine just below the crown were recovered from the early Middle Paleolithic of Sclayn, Belgium (Otte et al. 1985). These grooves looked as if they might have been made intentionally, in order to hang the teeth either on necklaces or as pendants. However, closer examination indicated that the grooves occurred naturally, probably due to grasses or other abrasives in the diet (Gautier 1986). (Similar grooves have been observed in hominin teeth, including specimens still in situ in the jaw (Formicola 1988; Ungar et al. 2003)).

Although there may be disagreements over individual specimens, there is probably little disagreement – in principle – that objects that can be explained taphonomically should be rejected as evidence for the elaboration of culture. However, taphonomic arguments have also been used in a rather different way, especially by Bednarik (1994).

His argument starts with a set of fairly obvious observations: that most art and most symbolic artifacts made at any one time period will never be recovered by archaeologists, that rock art will be preserved for long periods of time only if located in especially favorable locations such as deep caves, and that art made of bone or limestone will survive only in a basic or at least non-acidic environment. He then assumes that every person makes a certain number of symbolic artifacts, n, during a lifetime. It follows that if a population is small, if n is low, and if the recovery rate for symbolic artifacts is low, then the probability of any symbolic artifact finding its way into the hands of archaeologists is also low. This makes sense, and is something archaeologists need to keep in mind, but there are two reasons to be cautious about embracing Bednarik's argument too enthusiastically in the present context.

First, if taken literally, it would mean that we would infer that people were making symbolic artifacts in the absence of any evidence whatever, as long as people were making such artifacts at a later time. This amounts to an untestable hypothesis. Imagine two hypothetical situations:

1. A population at time t_1 (perhaps the late Middle Pleistocene) produced 10 symbolic artifacts per person per lifetime. The population was so small and the likelihood that any one artifact

would survive was so low that archaeologists recover no such artifacts. However, by time t_2 (perhaps the middle of the Upper
Pleistocene), population density had increased to the point where
symbolic artifacts show up regularly in the archaeological record. If, following Bednarik's reasoning, archaeologists inferred
that people at time t_1 were making symbolic artifacts, they would
be right.

2. Now imagine that at time t_1 people were making zero symbolic
artifacts per person per lifetime. The situation at time t_2 is the
same as that just described. The archaeological record would be
exactly the same – it would contain zero symbolic artifacts from
time t_1 and many symbolic artifacts from time t_2. Yet in this case,
inferring that people made such artifacts at time t_1 would be an
error.

If the test implications of a hypothesis and its null hypothesis are identical, then the hypothesis is untestable.

Second, it is not my purpose here to discover the very first origins of
symbolic behavior. It is, rather, to determine at what point culture was
elaborated into overarching systems in which virtually all behavior – and
all material culture – had symbolic cultural meaning and fulfilled a cultural as well as a practical function. In such circumstances, an n of 10, or
of 100, would be low. In one sense, almost 100% of artifacts made (other
than manufacturing by-products) would be symbolic or cultural, although
in many cases this cultural aspect would be invisible to archaeologists (as
I discuss shortly).

In addition, the scope of the inquiry is worldwide. If one accepts my
tentative conclusion in the previous chapter, that the biological prerequisites of culture had evolved at least by the end of the Middle Pleistocene,
then both the by-product and the anxiety hypotheses imply that the
elaboration of culture should have been universal by this time. In that
case, undeniable evidence of art, religion, ritual, and so forth should be,
if not abundant, at least not rare, and certainly not absent, for one simple
reason: in absolute terms the archaeological record for the late Middle
Pleistocene and early Upper Pleistocene is large and includes large quantities of bone and antler.

It is true that Neanderthals may have been a separate species who
had not evolved these capacities (but more on this later, too). In this case,
only *Homo sapiens* would have left behind evidence of elaborated culture, but this should be abundantly clear from the archaeological record
of that species. Moreover, the absence of such evidence from the Neanderthal archaeological record would reflect not low population density
and poor preservation but an absence of elaborated culture among Neanderthals.

5.1.3. Symbolic versus Practical Function

If plausible taphonomic explanations are to be preferred as simpler than cultural explanations, then the same is true of explanations that invoke a practical function rather than a symbolic, cultural meaning. A cut mark on an ungulate bone that can be explained by butchering will not be interpreted as symbolic or cultural in origin. This is *not* the same thing as saying that an artifact cannot have both a practical function and a cultural meaning. In the context of elaborated culture, most objects will have both. What I mean is simply that if an object or some aspect of an object can be interpreted as reflecting either a practical or a cultural explanation, then the former interpretation is simpler because we need not invoke either socially created coding or fully elaborated cultural systems. (Of course, as I stated earlier, if the argument behind a practical explanation is implausibly convoluted, then a cultural explanation is simpler.)

On the other hand, just because objects with obvious functions can be explained without recourse to the elaboration of culture, it does not follow that objects without obvious functions constitute good evidence for the elaboration of culture. It is archaeologists who are interpreting the archaeological record, and most of us have grown up in industrial societies rather than as hunter-gatherers using stone-age technology. Even in the Lower Pleistocene, making a living as a hunter-gatherer would have been a complex matter, and the technology was undoubtedly far more complex and varied than the stone tools alone. It is inconceivable to me that an archaeologist can look at an object and, with any real confidence, eliminate the possibility that it served some practical function.

Indeed, even if we were able to talk to a Neanderthal who lived in southern France 100,000 years ago, and we asked him or her to determine whether an object made in southern Africa at the same time had a practical function, we could not expect an accurate answer. The environmental differences involved and their effects on technology would be too great. Even a Neanderthal from southwestern France shown a contemporary artifact from southeastern France could not be relied on, because memetic traditions in the two areas might have been different. In short, only in extreme cases can one accept with any confidence the argument that some object must have been "non-utilitarian."

A few examples of objects that have been interpreted as non-utilitarian or symbolic may help to clarify these points. Symbolic meaning has been invoked to explain perforated bones, teeth, or stones, as well as notched, grooved, or serrated bones or stones (see especially Bednarik 1992; Marshack 1991), sometimes even when the excavators themselves have made no such claim. (Henshilwood and Sealy [1997:892] interpret serration on a bone point from Blombos Cave as "probably derived from binding to a haft."). Such objects include, for example, perforated bones from Pech de l'Azé II (Bordes 1969), Bocksteinschmiede (Wetzel and Bosinski 1969), and Repolusthöhle (Mottl 1951), serrated or edge-notched bones from Apollo Cave (Wendt 1974; 1976), Kebarah (Davis 1974), and the Klasies River Mouth (Singer and Wymer 1982:116).

Yet if one looks through a single ethnography, Murdoch's (1892/1988) *Ethnological Results of the Point Barrow Expedition,* one finds similar artifacts with practical functions:[*]

- Perforated wooden artifacts (Murdoch's figures 23–31, 108, 137–142, 205, 208–212, 215–221, 311–312)
- Perforated bone artifacts (figures 124, 128–129, 137–141, 193, 201, 234–236, 321)
- Perforated antler artifacts (figures 113–116, 118, 122)
- Perforated ivory artifacts (figures 158, 237, 247, 250, 208–212, 215–121, 311–312)
- Perforated stone artifacts (figures 108, 162–163, 226
- Grooved wooden artifacts (figures 157, 204)
- Grooved bone artifacts (figures 133, 172, 304)
- Grooved stone artifacts (figures 27, 28)
- A serrated bone artifact (figure 147)
- A serrated antler artifact (ff 323)

This certainly does not mean that we can determine the function of an object from the Lower Paleolithic of France or from the Middle Stone Age of South Africa by comparing it to an object of Eskimo material culture. What it does mean, however, is that there are practical reasons for perforating, grooving, notching, or serrating bone or stone. Thus an object that is perforated, grooved, notched, or serrated does not, ipso facto, constitute evidence for the elaboration of material culture, and in this review of the evidence I ignore claims where such features are the only basis for such an inference.

By the same token, many researchers have made claims for symbolism on the basis of scratches or cut marks, usually on bone, that seemed not to be related to butchering (e.g., Crémades 1996; Leonardi 1988; Raynal and Seguy 1986). (These are often described as "engravings," even when the mechanics of making the cuts is no different from that involved in butchering.) I have argued elsewhere (Chase 1995) that there

---

* In compiling this list, I ignored long grooves cut as hafting for blades, tools with metal components, barbs on projectile points or harpoons, elaborate objects such as boxes and hide scraper handles, and purely symbolic or decorative objects such as masks or bead, and toys.

are other ways of explaining such marks without recourse to symbolism, including the use of bones (or stones) as cutting boards.

What is needed, at a minimum, to infer a symbolic meaning for such cuts or scratches is evidence that the person who made them intended to create a visually meaningful pattern of marks. I believe this also applies to marks other than simple cuts or scratches. For example, Bednarik (1992:38) cites a fragment of antler from the Middle Paleolithic of the Grotte Vaufrey in France as evidence of symbolism, even though "the notches are of different sizes and unevenly spaced." The archaeologist who described the piece cited it only as evidence that the inhabitants of the site understood the material and how to work it (Vincent 1988:191).

This argument applies not only to bone but also to other materials. For example, a fragment of ostrich eggshell from Apollo Cave (Vogelsang 1998: 81) and a Levallois core from Qafzeh (Hovers et al. 1997) bear sets of cut marks. The materials are somewhat unusual, and neither may have been used as a cutting board, but there is no evidence that the marks were intended either to serve as symbols or to create a visually meaningful pattern.

In some cases, archaeologists have inferred an intention to create a visually meaningful pattern of marks where, I believe, there is a simpler explanation. The best example is the apparent patterning of cut marks on four bone fragments from the Middle Pleistocene of Bilzingsleben, Germany (Mania and Mania 1988) There is no question that the marks on at least two of these bones are striking, but an extraordinary number of cut marks appear on the large animal bones from the site (see illustrations in Mania and Weber 1986), and for the great majority of them no one has suggested a symbolic or cultural explanation (but see Behm-Blancke 1987). It is not surprising, therefore, that some striking patterns would appear by chance alone.

Moreover, many of the marks were probably made in ways that produced visually striking patterns, but without any intention of doing so. For example, if a bone were used as a cutting board, it is quite plausible that a series of parallel or subparallel cuts would be made. My own cutting board at home has such marks from slicing bread and dicing vegetables. In evaluating the importance of eye-catching pieces from one end of a spectrum of cut-marked bone, one must consider the rest of the assemblage.[*]

---

[*] Although our interpretations of the cut marks from Bilzingsleben may not be exactly the same, I wish to thank the Manias for their hospitality and openness when I visited the site. This is another example of an honest disagreement based on the inevitable subjectivity involved in interpreting the archaeological record.

In my review of the evidence, I consider only pieces for which a reasonable case can be made that the maker intended to create a visually meaningful pattern.

5.1.4. Culture versus Curiosity

For the problem being addressed here, the only relevant data are those that constitute evidence for the elaboration of culture. My purpose in this chapter is not to pinpoint the origin, among hominins, of the capacity for symbolism or even of its use. Nor is it to decide when our ancestors first developed an aesthetic sensibility. Still less is it to determine at what point hominins became capable of recognizing iconic resemblances between an object and either an animal or a person. It is certainly not to find evidence for the origins of curiosity.

For this reason, two classes of data that have, rightly, attracted attention in the archaeological literature are irrelevant here. The first is manuports, objects that are unusual in appearance or that, at least to modern archaeologists' eyes, are iconic in nature, that have been brought to an archaeological site from elsewhere, but that apparently were neither modified nor used for practical purposes. These include a pebble from Makapansgat, South Africa, that bears some resemblance to a face (Bednarik 1998; Dart 1974). The second class of data consists of unusual materials brought into a site and apparently used for practical purposes, such as one of the six quartz crystals from the Acheulian of Singi Talav, India (D'Errico et al. 1989). Some of these include normal tools made of unusual materials. For such artifacts, at least three explanations are possible:

1. Unusual materials were used to make ritual or symbolic artifacts that otherwise resemble ordinary functional artifacts.
2. Unusual materials were collected and, after their novelty had worn off, were used to make ordinary artifacts for practical uses.
3. No notice was taken of the unusual nature of the material. It was used, like any commonplace material, to make ordinary artifacts or to serve ordinary purposes.

Unless the last two explanations can be eliminated, such objects do not constitute evidence for the elaboration of culture into overarching symbolic systems.

For example, Oakley (1971; 1973) reported four unusual objects, all of them probably of Acheulian age. A biface was collected in 1911 at West Tofts, Norfolk, England. Although the entire margin had been bifacially flaked, a portion of the cortex remained on one face, with a fossilized bivalve, *Spondylus spinosus,* clearly visible in the middle of it. Oakley argued that the maker of the biface probably worked carefully around the fossil. But even if so, no sacrifice in terms of shaping or sharpening was necessary; the core has been worked into a typical biface. It is equally plausible that the fossil made no impression whatever on the flintknapper, who simply treated the nodule like any other core.

On a similar, rather crude biface from the Middle Gravels of Swanscombe, flaking removed part of an embedded *Conulus* fossil (Oakley 1981:210-211). At Saint-Juste-des-Marais, Oise, France, an echinoid (*Micraster*) fossilized in flint was retouched into a scraper. It was recovered from river gravel associated with an Acheulian industry. Two flakes were recovered from the Middle Gravel at Swanscombe, England. They were of *Isastraea* chert, that is, of fossilized coral, which is allocthonous in the Thames catchment.

It is possible that these fossils had cultural meaning for the Acheulian people who made the scraper and the flakes, but it is equally possible that the flint and chert were used like any other flint or chert, without regard to the fossils. (Oakley cited cultural meanings assigned to echinoid and coral fossils in historic folklore, but given the huge amount of time that has passed since the Acheulian, it makes little sense to argue for cultural continuity of this kind.) Moreover, even if the makers of these artifacts noticed and cared about the fossils, it does not necessarily follow that any of them had symbolic cultural meaning. They may merely have been regarded as curiosities or aesthetically pleasing objects.

5.1.5. The Problem of Equivocal Evidence

It is inevitable that some evidence will be equivocal, in the sense that it is impossible to invalidate either an explanation for it involving the elaboration of culture or a simpler explanation that invokes only taphonomy, practical function, or curiosity. Actually, since in archaeology little is absolutely certain, such items are ones for which either kind of explanation is plausible – that is, neither is highly improbable. I have mentioned one such case, the flutelike bear bone from Divje Babe. There is a certain subjectivity to all such cases, because different archaeologists evaluate relative probabilities differently. Some archaeologists (myself included) find a taphonomic explanation of the flutelike specimen more probable than a musical one, whereas others find it less probable. This particular specimen dates to about 43,000 years ago (Turk 1997b) and is therefore too recent to be relevant to testing our hypotheses in any case (for reasons discussed shortly). However, I will discuss several other equivocal specimens later.

In discussing how such data were to be interpreted when documenting the origins of symbolism, Dibble and I (Chase and Dibble 1992) argued that equivocal evidence should be accepted only when the plausibility of other explanations was very low. I continue to believe that this stand is correct as a criterion for documenting the origins of a phenomenon such as the use of symbols. That is why, in the previous chapter, I said that the evidence for the origins of socially constructed coding was suggestive but not conclusive. In the context of the issues facing us in this chapter, however, the logic of this argument requires closer examination.

The dilemma posed by equivocal evidence is similar to one that is explicitly recognized in statistics. Statisticians refer to accepting a false hypothesis as a type I error, and to rejecting a true hypothesis as a type II error. By convention, in statistics, a hypothesis is accepted when the probability of making a type I error is less than 1 in 20 ($p < 0.05$).

The logic of my argument is similar, but not exactly congruent. In statistical analysis, a single hypothesis is tested against a null hypothesis, which states that a random sample may misrepresent the population from which it was drawn to such an extent that the analyst is misled into accepting a false hypothesis.[*] The generally accepted (statistical and scientific) attitude is that a hypothesis should be considered as having been confirmed only if the probability that the null hypothesis is true is quite small.

In the archaeological cases Dibble and I were evaluating, the hypothesis was that a given population was using symbolism. The null hypothesis was that taphonomic or other factors had produced what looked like evidence of symbolism. Thus the null hypothesis was based not on sampling error but on equifinality. We argued, using the same philosophy that underlies statistical analysis, that the null hypothesis should be rejected only when it was highly implausible. To argue the contrary, that one should infer the use of symbolism when the validity of the null hypothesis is probable, would be equivalent to accepting a statistical test even when p is very high.

In the present context, however, there is an additional consideration. We are not only testing individual hypotheses against a null hypothesis (the hypothesis of equifinality), but we are testing non-null hypotheses against each other as well. We are using archaeological evidence to choose between two sets of hypotheses:

- On the one hand, the group benefit hypothesis (H_{GB}). This hypothesis will be rejected if good evidence of cultural elaboration is found at the beginning of the Upper Pleistocene.

---

[*] For example, the mean of the sample may deviate from the true mean of the population (a possibility t tests are designed to evaluate). Likewise, the proportions of different categories in a sample may deviate from those in the population (a possibility addressed by the chi-squared test).

- On the other hand, the by-product hypothesis (H_B) and the anxiety hypothesis (H_A). These hypotheses will be rejected if no good evidence of elaborated culture is found at the beginning of the Upper Pleistocene.

This changes the logic of the procedure.

To clarify this point, consider the first column in Table 5.1.

Table 5.1. Consequences of erroneously accepting or rejecting evidence for cultural elaboration.

	H_1	H_{GB}	H_B & H_A
Erroneously *accept* evdence for cultural elaboration	Type I error	Type I error	Type II error
Erroneously *reject* evidence for cultural elaboration	Type II error	Type II error	Type I error

In this case, a single hypothesis (H_1), that culture was elaborated at time *t,* is tested against the null hypothesis (column 1). The consequences of erroneously accepting or rejecting archaeological data as evidence for elaboration are clear. Setting a high standard of evidence (the equivalent of a low value for *p*) minimizes the probability of making a type I error. This is identical, logically, to the argument Dibble and I made with regard to symbolism.

However, in the present case, the consequences are not so simple. If we set a high standard of evidence, the consequences for H_{GB} and for H_A and H_B are diametrically opposed, as shown in the second and third columns of Table 5.1. Setting high standards of evidence for cultural elaboration minimizes the probability of a type I error with regard to H_{GB}. The probability of making a type I error with regard to H_B and H_A, however, is minimized by setting a *low* standard of evidence. We thus are faced with two competing demands: making allowances for equifinality and trying not to bias the case against the anxiety and by-product hypotheses. It is clear that in this case we must be more willing to accept equivocal evidence, or at least to consider it, than if we were considering a single hypothesis.

The best solution would be to find a different test implication or set of test implications. Within archaeology, I cannot think what such a test implication would be, although it should be possible to approach the problem using data from another discipline. Otherwise, I know of no generally accepted procedure for balancing these two demands. Fortunately, there is an additional aspect of the test implications for these three hypotheses that will help to minimize the effects of this dilemma. It has to do with the predicted temporal and spatial distribution of evidence for cultural elaboration.

5.1.6. Questions of Time and Space

In the previous chapter, I accepted that the data suggested, tentatively, that socially created coding was part of the hominin adaptation by the end of the Middle Pleistocene. The question now is whether or not, by the end of or immediately following the Middle Pleistocene, the elaboration of culture had shaped the archaeological record of all hominins, or (if Neanderthals were a separate species) of all *Homo sapiens.*

This means that the time periods of interest to us are the Middle Pleistocene itself and the early Upper Pleistocene. Whether there was symbolism in the later Middle Paleolithic, or Middle Stone Age is not an issue, because that has been granted as part of my working premise. Nor, for the moment, does it matter whether Neanderthals were using symbols. We want to know only whether elaborated culture appeared simultaneously with socially created coding or whether there was a significant time lag.

This has two implications for the way we approach the data. First, if socially created coding dates to the end of the Middle Pleistocene or before, then we are not interested in what happened in the middle of the Upper Pleistocene or later. I review only evidence from the Middle Pleistocene or earlier and from the first part of the Upper Pleistocene, that is, from oxygen isotope stage 5. An absence of evidence for cultural elaboration before stage 4 would mean a lapse of more than 50,000 years between the origins of socially constructed coding and the elaboration of culture. The anxiety and by-product hypotheses would have to be rejected.

Second, as I pointed out in chapter 3, the group benefit hypothesis is consistent with a gradual, sporadic, on-and-off appearance of the elaboration of culture. The other two hypotheses are not. This means that we must reject the by-product hypothesis unless we find either of the following:

- An abrupt, decisive, and widespread appearance of evidence for the elaboration of culture
- The spread of such evidence from a single center in a manner that could be reasonably correlated with the spread of a population (such as modern *Homo sapiens*) that had evolved the neural capacity for socially created coding

Scarce or sporadic evidence (equivocal or not) of the kind indicative of elaborated culture will not suffice to confirm either the anxiety or the by-product hypothesis.

With these criteria in mind, we can now turn to a consideration of the archaeological record itself.

5.2. THE ARCHAEOLOGICAL EVIDENCE

5.2.1. Mortuary Practices

A lively debate has taken place in archaeology and human paleontology about whether or not Neanderthals buried their dead (e.g., Belfer-Cohen and Hovers 1992; Binant 1991; Bocquet-Appel and Arsuaga 2001; Bocquet-Appel and Arsuaga 1999; Bonifay 1989; Carbonell et al. 2003; Dibble and Chase 1993; Farizy 1988; Gargett 1989; 1999; 2000; Harrold 1980; Hovers et al. 2000; Smirnov 1989; Tillier 1995; Tillier et al. 1988; Watson 1970). Claims and counterclaims have been made concerning ritual, including grave goods, cannibalism, and secondary burial prior to the Upper Paleolithic (e.g., Bar-Yosef et al. 1988; Breuil and Obermaier 1909; Clark et al. 2003; Defleur et al. 1993; Defleur et al. 1999; Fernández-Jalvo et al. 1999; Le Mort 1987; 1989b; 1989a; Leroi-Gourhan 1975; 1998; Lietava 1992; Russell 1987b; 1987a; Solecki 1975; 1977; Sommer 1999; Stiner 1991; Tappen 1987; Tillier et al. 1991; Toth and White 1990-1991; Trinkaus 1985; Vandermeersch 1970; Villa 1992; Villa et al. 1988; White 1986; 1987; 1995; White et al. 2003; White and Toth 1991). In a general context, these claims and this debate are of real interest. In the present context, they are irrelevant. If we have accepted that socially constructed coding appeared long before the Upper Paleolithic, then the question of whether or not people were capable of symbolic mortuary behavior before this time is not in question.

There is nothing in the group benefit hypothesis that excludes the possibility that symbolism was used in a limited way to alleviate emotional distress before socially created coding was elaborated into systems that included virtually *all* of what humans do, perceive, or think. It is this elaboration that I am trying to explain and that the anxiety hypothesis is designed to explain. In other words, the anxiety hypothesis states that the all-encompassing cultural systems that are universal among humans today were created to allay individuals' anxieties. It thus goes far beyond stating merely that socially created symbolic coding may be used, or may in the past have been used in a more limited manner, to allay anxieties (or other strong emotions such as grief).

Because death certainly evokes emotions such as fear and grief, it is perfectly possible that, once evolution of the brain made symbolism and socially created coding available to humans, they might have used them to allay these emotions. It does not necessarily follow that in order to do so they elaborated symbolism and socially created coding into overarching, all-encompassing cultural systems. It is this that the anxiety hypothesis postulates. If the anxiety hypothesis is correct, then the use of culture in a mortuary context would have to be accompanied, simultaneously, by fully elaborated culture. All aspects of the archaeological record, not just those related to death, should show the effects of culture. Thus, the anxiety hypothesis must be supported by evidence unrelated to death and burial. In the context of this chapter, then, there is no reason to consider evidence related to mortuary practices.

5.2.2. Standardization and Style

Two closely related concepts, style and standardization, might seem to provide a way to determine whether the morphologies of stone tools or other artifacts were culturally determined, or at least culturally influenced. I believe, however, that neither is a valid basis for making such a decision. I will tackle the question of style first, because it will lay the basis for discussing the question of standardization.

A lively debate about the nature of style has taken place in the archaeological literature (Byers 1994; 1999; 2001; Clark 1989; Close 1980; 1989; Conkey 1978; 1990; Dunnell 1978; Falk 1980b; Jelinek 1976; Sackett 1973; 1982; 1985; 1986; Stiles 1979; Wiessner 1983; 1984; 1990; Wobst 1977). Some aspects of this debate are irrelevant to the task at hand, but the concept of style is useful because it puts standardization into the context of two crucial factors – the overdetermination of artifact form, and the patterning of similarity and differences of artifact form through time and space.

Two themes seem to be common to most of the definitions of style in the Paleolithic archaeological literature:

- Style is thought of as something that is associated (for whatever reason) with a given group of people within a given geographical range over a given span of time. As Conkey (1990:6) put it, "if there is any one thing we have had in the backs of our minds for the use of stylistic analysis, it has been to find or reveal social units or specific historical entities." How a social or historical group is to be defined is often left unstated, but from an archaeological perspective, style is usually considered to be an index of a historically and ethnically or culturally bounded social unit

(Conkey 1980; 1990; Dunnell 1978; Jelinek 1976; Sackett 1973; 1982; 1985; Stiles 1979; Wiessner 1983; 1984; 1990).

- Style is generally considered to be, in Byers's words (1994:379), "an overdetermination of form with respect to end-goal requirements." Such end-goal requirements include the practical use to which a tool will be put and the technological requirements of its manufacture. The notion of overdetermination is included even in what Sackett (1982) called "isochrestic" style, that is, choices made in the process of manufacture among procedures that, from a functional or technological perspective, are equally valid. In this case, it is isochrestic choices rather than "adjunct" decoration that overdetermine the form of the artifact, but the overdetermination still exists. The concept of overdetermination is also invoked by a school of thought that holds style to be something left over after the requirements placed on artifact form by function, technology, and the like (Close 1980; 1989; Dunnell 1978; Jelinek 1976). From all these perspectives, the presence of style means that artifact morphology does not vary as freely as function and technology would permit.

I therefore use the word "style" to mean overdetermination of artifact form that is, in one way or another, associated with a given group of people bounded both ethnically and temporally.

The archaeologist must ask three questions, however, before he or she can use the presence of style in the archaeological record to infer the presence of any other phenomenon:

1. Is my definition of style valid? Is it to be expected that overdetermination of artifact form will map spatially and temporally with different ethnic or historically bounded social groups? There is no question that practical considerations make the link between style and historically or ethnically bounded groups less clear than the preceding summary implies (Jelinek 1976). Style may be used to convey information about something other than group identity (Wiessner 1983; 1984; 1990; Wobst 1977); not all ethnic boundaries are correlated with stylistic boundaries (Wiessner 1983); and style may be associated with ideas that cross-cut stylistic boundaries. These considerations are relevant to archaeologists using style as an index of prehistoric social groups. They are not a problem in the present context. Styles that reflect cultural boundaries rather than ethnic boundaries, or that are associated with cultural ideas rather than social groups, are still the products of elaborated culture and will still produce the same pattern of spatial and temporal uniformity and variability.

2. Can the archaeologist distinguish between similarities and differences due to overdetermination and those due to other causes such as practical function and the demands of raw material? It is extremely difficult in practice to distinguish between morphological characteristics that are determined by the demands of raw material or function and those that represent overdetermination (see Chase 1991). Are the basal thinning of Emireh points and the tangs of Aterian artifacts related to hafting? Were the backed microliths of the Howieson's Poort made to be mounted in compound tools? In addition, the archaeological record may contain artifacts whose morphology represents not a desired end product but the processes of breaking, resharpening, and so forth, to the point where the tool becomes useless (Dibble 1987; 1989; 1995; Frison 1968; Jelinek 1976). For any given class of artifacts, it must be demonstrated that such practical problems can be overcome.

3. Assuming questions 1 and 2 can or could be answered positively, then what social or cultural phenomena produce this kind of overdetermination? If anything other than elaborated culture can produce stylistic patterning of the archaeological record, then the archaeologist cannot infer the elaboration of culture from the presence of style. It is this question that I address now.

Generally speaking, two mechanisms are cited by which a style or styles are associated with a given group of people:

•   Style may be a means of conveying information, including, especially, information about ethnic or personal identity (Conkey 1978; 1980; Wiessner 1983; 1984; 1990; Wobst 1977).

•   Style may be the result, intentional or not, of adhering to group standards and norms concerning the manufacture of stone tools (Byers 1994; 1999; 2001; Conkey 1978; 1980; Sackett 1973; 1982; 1985; 1986; Stiles 1979; Wiessner 1983; 1984; 1990; Wobst 1977).

The first mechanism clearly implies a context of elaborated culture. All monkeys and apes have personal and social identities of which they and others are aware (male-female, young-old, parent-child, ally-rival, friend-stranger, etc.). They use these as the basis of a complex social life, yet they do so without using material culture to convey information about those identities. This is possible because such identities do not need to be marked symbolically; they can be learned by observation and experience.

When socially constructed coding is elaborated into all-encompassing cultural systems, however, many identities are culturally created and culturally defined – godparent, spouse, and citizen, for example, or clan, tribe, and nation. It is natural that such identities are marked symbolically by body decoration, style, and the like, because they are no longer self-evident; they cannot be recognized by observation alone. In the process, "natural" social identities become, like everything else, enmeshed in the web of cultural significance, and they are likely to be symbolically marked as well. In this sense, it is altogether normal that elaborated culture will be reflected in the archaeological record by style.

The second explanation is more problematic. Even if style is not used deliberately to convey information, adherence to cultural standards concerning artifact manufacture may well produce what Sackett (1986) has called "passive style." If culture provides norms concerning either how artifacts are made or what they should look like – norms that are more restrictive than the demands of function and raw material – then there will be an overdetermination of form. And if these norms vary from one culture to another or change through time as cultures themselves change, then artifact forms will vary as well. The result will be a patterning of the archaeological record that must be recognized as style (as I have defined it).

The problem, lies with the concept of "norm" or "standard." Byers (1994; 2001) states explicitly what other proponents of this position seem to assume implicitly: that the standards that produce style are cultural. Certainly, norms and standards regarding the making or morphology of stone tools are common, if not inevitable, in the context of fully elaborated culture. Thus, there can be no doubt that adherence to cultural norms and standards will produce overdetermination of artifact form, and that the archaeological patterning produced by this overdetermination will vary through time and across space, reflecting the distributions of the underlying cultures.

It follows from either of the preceding cultural explanations for the creation of style that style will be observed in the archaeological record as a pattern of variation in stone artifact morphology that is discontinuous through space and time (see Stiles 1979:5). Any one geographic area will, at any given time, be stylistically homogenous (unless occupied by more than one ethnic group). That is, the appearance of each category of artifact will be overdetermined in the same manner (in the same style).

At the same time, there will be disjunctions in style both across space and through time; adjacent regions or sequential time periods should be marked by different styles. These similarities and disjunctions, by definition, are not attributable to changing conditions that demand new technologies or new tool forms for purely practical purposes. The reason is that similarities and differences in style reflect similarities and differences in culture. Culture, being based on arbitrary symbolism, tends to vary from group to group and to change through time. If culture produces style, then styles will vary from group to group and will change through time as well.

One must, ask, however, whether memetic traditions, which owe nothing to elaborated culture, may not produce similar patterns of over-determination (see McNabb et al. 2004). Style is observable among primates who do not make use of symbols and whose behavior is not constrained or guided by cultural norms but is guided by memes common in each group. For example, different groups of chimpanzees have different methods of making termite fishing wands and breaking open nuts, different postures when being groomed, and so forth (Whiten et al. 1999). From the perspective of technology, perhaps the most striking example comes from the work of McGrew, Tutin, and Baldwin (1979), who found that some chimpanzee groups peeled the bark off twigs used for termite fishing, whereas others did not, apparently with no effect on function. This is clearly a stylistic difference, yet it is also clearly memetic rather than symbolic.

Just how applicable such an example is to the hominin archaeological record is problematic. Certainly, in some cases there is little doubt that cultural standards underlie patterns of systematic variation in artifact form through time and space. Few would argue, for example, that this was not the case in the Upper Paleolithic of Europe – certainly not in the later Upper Paleolithic. Our purpose, however, is not to understand the Upper Paleolithic but to try to identify the elaboration of socially constructed coding into fully elaborated culture. If we look at patterns of stylistic variation before the Upper Paleolithic, their cause is not so clear.

A number of scholars have recognized that the Mousterian industries of the Levant seem to fall into several categories. The differences among them do not involve shaping flakes by retouch, but rather the shapes of the raw flakes and the methods by which the flakes were removed from the cores (Copeland 1975; Jelinek 1981; Meignen and Bar-Yosef 1992):

> 1. The first group includes assemblages which display mainly broad, radially prepared flakes. ...
> 2. The second group includes assemblages with elongated blanks (blades, subtriangular blades and points) which were produced mostly by recurrent unidirectional methods (often converging). ...
> 3. The third group is characterized by the production of short blanks (flakes and short, broad-based points) that are mostly obtained through unidirectional recurrent methods (often converging), but also by radial reduction. (Meignen and Bar-Yosef 1992:143)

It is not clear whether the methods of core reduction were intended to produce flakes of a given shape or whether flake shape was a by-product of the chosen method of core reduction. The important point is that these different kinds of assemblages are found at several sites within a limited region and replaced one another through time. In these respects they fit the criterion of style.

Other phenomena in the Middle Paleolithic or Middle Stone Age raise the same question. There are differences in flaking technologies and in core reduction strategies. There are a few types such as Emireh points and Prodniks that are limited in both time and space. There are ways of making stone tools, such as the tangs on Aterian tools and the backing characteristic of the Howieson's Poort, that are also limited in time and space.

Let us assume for the moment that such differences among assemblages represent overdetermination of form, that they are not the products of different functions associated with different adaptive strategies. In this case, all these examples would fit the definition of style given earlier. The question remains, however, whether these differences reflect different symbolic cultural standards or simply different memes. That is, youngsters learning to work stone might have learned the methods used by their elders not because of any cultural conventions but simply because flintknapping at this level of sophistication is not simple, and learning from available models would have been more efficient than trying out a random variety of strategies.

It is my opinion that we simply cannot answer this question. Although geographically limited memetic traditions can be observed among nonhuman primates today, we cannot observe the spread or persistence of such traditions on a geological time scale. The fact is, quite simply, that no population of technologically sophisticated, acultural stone tool makers exists today. We cannot observe their behavior even over the space of days or years, much less over millennia. One archaeologist might believe that memetic traditions were so ephemeral that they are not reflected in the archaeological record. Another might believe that memetic traditions simply did not exist, because, in the absence of cultural norms, individuals would have developed idiosyncratic methods of knapping. A third might believe that memetic traditions are a perfectly plausible explanation for the kinds of phenomena observed in Middle Paleolithic stone tools.

The problem is that we have no objective way of choosing among these opinions. We do not really know how long a memetic tradition can last among monkeys or apes. We certainly do not know how long a memetic tradition might have lasted among *Homo ergaster* or Neanderthals or even among anatomically modern but acultural *Homo sapiens*. Because social as well as cognitive factors would have been involved, I am highly skeptical of psychology's ability to help us in this regard. Our ideas on the subject consequently amount to no more than opinions. No matter how well thought out or how thoroughly grounded theoretically, they will always, I fear, be untestable and therefore fundamentally subjective.

If there is no practical way to eliminate the possibility that memetic traditions, rather than cultural norms, produced stylistic variation in the archaeological record, then stylistic variation cannot be used to infer the presence of elaborated culture – the possibility of equifinality cannot be ignored. Style cannot be a valid indicator of the elaboration of culture unless (1) a way can be found to distinguish meme-based style from culture-based style, or (2) memetic traditions can somehow be eliminated as a source of stylistic variation.

Let us now turn to the related question of artifact standardization that is uniform rather than variable through time and over space. Holloway attributed such standardization to the presence of what I would call elaborated culture:

> The important question is whether or not other processes were also operating, such as consensus, or explicit rules about the forming processes. What is at issue is concatenated activity according to rules, i.e., grammar. Imitation and observational learning seem to me insufficient to explain the tremendous time depth and wide geographic extension of certain tool types in much of the Old World. It seems more likely that rules, consensus, syntax did exist, and that a communication system using symbolic language existed at least by the time of handaxes, if not before. (Holloway 1969:401)

Other archaeologists seem, in one way or another, to echo this point of view, or at least to see standardization (or the related concepts of "mental template" and "imposition of form") as linked, in some way, to arbitrary cultural standards (e.g., Dibble 1989; Gowlett 1984b; 1996; Mellars 1989; 1996a:133-136; Monnier 2000; Nowell 2000; Wurz 1999; see Marks et al. 2001 for a review). I have probably contributed to this perception (Chase 1991). However, the theoretical underpinnings of this perspective need to be examined in light of the conceptual differentiation between memetic and fully elaborated human culture.

It is always dangerous to try to read the minds of others, but I suspect that one reason standardization has been seen as a correlate of symbolism and culture is that it is, in fact, one element of style. Style consists of overdetermination of form that is correlated with an ethnically and historically bounded social group, and overdetermination is, of course, standardization. Obviously, then, wherever there is stylistic patterning, there will be standardization. Either using style to communicate or adhering to cultural norms will overdetermine artifact forms, and within a limited block of time and space this overdetermination will be uniform, at least for each artifact type – all hide scrapers, for example, will resemble each other. Standardization must be present for there to be style.

It does not follow, however, that all standardization reflects either style or cultural norms. Without further bridging arguments, neither careful workmanship nor standardization of artifact morphology or production indicates, in itself, the presence of cultural norms or standards (Chase 1991). Indeed, there is good reason to believe that the kind of standardization referred to by Holloway has no relationship to culture.

The fine working edges and symmetrical forms of at least a minority of Acheulean bifaces have been long known. Gowlett (1984b) discovered that bifaces from Kilombe Falls, Kenya, were standardized in an unexpected way. Their length-to-width ratios were remarkably uniform (or at least length and width were highly correlated). The same sort of standardization has been found at a wide range of Acheulian sites (Alimen and Vignal 1952; Dibble 1989; McPherron 1994; 1995; 2000). In addition, a regularized allometry is involved: the length-to-width ratio shifts in a regular manner depending on the overall size of the biface (Crompton and Gowlett 1993; Gowlett 1996; Gowlett and Crompton 1994; McPherron 1994; 1995; 2000). Gowlett argued that these regularities were not imposed by functional or technological constraints, which would permit a greater variability than the actual distributions of the ratios. He concluded that these regularities indicated that the bifaces had been made according to a set of "instruction sets" or "rule systems," and he suggested that these were "culturally" maintained.[*]

The implications of this regularity are in fact exactly the opposite. Very similar length-to-width relationships have been found in virtually all Acheulian sites where they have been studied. Wynn and Tierson (1990) argued for regional differences, but this was refuted by McPherron (2000). Such uniformity is exactly the reverse of the pattern to be expected from culture-based style.

There are three parts to stylistic variation: overdetermination, uniformity within a restricted area and time period, and variation across space and time. The regularities found by Gowlett are observable across continents and over a vast span of time. This universality implies that some technological constraint was at work (Dibble 1989; McPherron 2000). It precludes a cultural basis, and probably a memetic basis as well. It is inconceivable that a single cultural norm for the making of handaxes remained unchanged for over a million years and over enormous areas of the Old World. The same can be said of Holloway's argument. The very time depth and geographic extension of the types to

---

[*] By "culture," Gowlett seems to have meant what I call "memetic tradition."

which he referred indicate that their uniformity had nothing to do with culture.

5.2.3. Isolated Artifacts that May Be Symbolic or Cultural in Nature

Individual objects whose morphology, decoration, or other attributes indicate that they served some symbolic purpose are a much more relevant class of evidence for our purposes. The symbolism involved need not have been referential. Equally important would be the kinds of cultural meaning usually associated with decoration or style. Although isolated finds of such objects do not constitute evidence for the full elaboration of culture into all-encompassing systems, people whose behavior was governed by such a system would inevitably have produced such artifacts.

There are objects from before oxygen isotope stage 4 that fit or have been claimed to fit this description. Some of them do not meet the criteria I discussed earlier, but others merit consideration.

In upper Bed I of Olduvai Gorge, Tanzania, a small (7.9 x 5.4 x 4.9 cm) cobble was found that had been pecked over most of its surface (Leakey 1971:84, 269, plate 18). It had an almost continuous artificial groove around one surface, as well as a series of small pecked depressions (Figure 5.1). Experiments showed that a cord could be tied around the groove. Leakey felt that it was "unlikely that it could have served as a tool for any practical purpose." She went on to say that

> a great deal of imagination is required in order to see any pattern or significance in the form. With oblique lighting, however, there is a suggestion of an elongate, baboon-like muzzle with faint indications of a mouth and nostrils. By what is probably no more than a coincidence, the pecked groove on the Olduvai stone is reproduced on the Makapansgat specimen by a similar but natural groove and in both specimens the positions of the grooves correspond to what would be the base of the hair line if an anthropomorphic interpretation is considered. (Leakey 1971:269)

It seems to me that this specimen indicates how easy it is let one's imagination produce a more complex explanation than is warranted. In this case, it is much more probable that this cobble had some practical use than that it was a work of art – even if we archaeologists, who have never made our living as Oldowan people did, cannot recognize its function.

In Auditorium Cave, south of Bhopal, India, Bednarik (2001; 2002:92) reported a boulder with a "cupule" and "a circuitous, hammered line that wraps around part of the cupule's perimeter. ... Both petroglyphs were clearly produced by impact and are as weathered as the surrounding rock surface.[*] They were buried about 30 cm below a Middle Paleolithic breccia, within an Acheulian level; no further dating is available for this level. The meaning of these phenomena is unclear. Although the cupules might possibly have served some utilitarian purpose, it is more difficult, on the basis of Bednarik's illustrations, to make that argument for the two grooves. Still, I find it difficult to accept this discovery as conclusive, rather than possible, evidence of symbolic intent and elaborated culture.

Figure 5.1. Pecked cobble from Olduvai Gorge, Tanzania. (Photograph by permission of Cambridge University Press)

A visually striking object was recovered from a road cut near the town of Tan-Tan, Morocco, in a level associated with Acheulean bifaces (Bednarik 2003; Kuckenburg 2001). On the basis of the morphology of the bifaces, this level was tentatively dated to 300,000 to 500,000 years ago.

---

[*] I do not provide illustrations for Auditorium Cave or for Tan-Tan, because Bednarik would grant permission to use them only on condition that he approve my interpretations of them.

The object is a small piece of metamorphosed quartzite, 5.8 x 2.6 x 1.2 cm in size, that to modern eyes bears some resemblance to a stylized human. However, the overall form of the object is entirely natural. The resemblance to a human is created by natural grooves running both longitudinally and transversely. Within five of the transverse grooves, Bednarik observed traces that he was able to replicate with indirect percussion using silica flakes and a hammer stone. According to his description, five of the eight transverse grooves on the upper and lower surfaces had been enhanced (although it is not entirely clear to just what extent) by percussion. Bednarik also observed minute particles of iron and manganese that resembled "manually applied paint residues" on Upper Paleolithic stone objects. He did not observe similar inclusions on other objects from Tan-Tan, although some did bear "an incipient ferromanganeous film, probably a 'fossil' rock varnish" (Bednarik 2003:405).

It is hard to know exactly how to interpret this object. Because it was probably a manuport, and because some grooves had been artificially emphasized, Bednarik argued that it demonstrated that Acheulian peoples had the capacity to recognize iconic resemblances of stone objects to human forms. This may be only partially true. It is certainly plausible that, visually, the object struck some chord for them, but whether they perceived a resemblance to a human form is less clear. In the present context, however, this question is unimportant.

What is important is whether this object constitutes solid evidence for elaborated culture. It does not, even if we accept Bednarik's argument that its iconic resemblance to a human figure was perceived and enhanced. There is nothing about its form that indicates that it was created to fulfill a symbolic cultural function, rather than simply serving as an iconic curiosity. If the specks of red that Bednarik observed are in fact remnants of paint, then the odds in favor of a symbolic interpretation would be improved. Nevertheless, the overall significance of the object is uncertain. I would place it in the category of equivocal evidence as an object whose cultural nature can be neither demonstrated nor disproven.

One of the most intriguing finds in prehistory is a small (35 x 25 x 21 mm) lapilli tuff pebble bearing human-made grooves from the Acheulian site of Berekhat Ram in the Golan Heights (Goren-Inbar 1986) (Figure 5.2). It came from a level underlying a 233,000 ± 3,000-year-old basalt flow (Feraud et al. 1983). Although there was initially some question about whether the grooves were natural or not (Goren-Inbar and Peltz 1995; Marshack 1997; Pelcin 1994), a careful study by D'Errico and Nowell (2000) confirmed that someone had cut several grooves into the pebble.

Goren-Inbar interpreted this object as a female figurine: "It seems that the occupants of the site selected a pebble that bore some characteristics of a female body. These were enhanced by adding incised grooves delimiting the head and arms" (Goren-Inbar 1986:11). Marshack (1997) carried this interpretation much further. The reading, however, is not self-evident. The resemblance to a human is much less clear than that of the Tan-Tan specimen. Indeed, judging from photographs, from most angles the Berekhat Ram specimen resembles a human only in that it has a knob that can be interpreted as a head.

Figure 5.2. Grooved pebble from Berekhat Ram. Bar is 1 cm (Photo courtesy of Cambridge University Press and Francesco D'Errico)

When viewed in "profile," however, it does bear a certain resemblance to some of the more globular, highly stylized female figurines of the European Upper Paleolithic. I cannot read either Goren-Inbar's or Marshack's mind, but I suspect that it was this latter resemblance that struck them, rather than any direct resemblance to a human figure (Noble and Davidson 1996:75-76; see also D'Errico and Nowell 2000). In Goren-Inbar's words, "it is only natural that the grooved item discovered be identified as a 'figurine'. Ignoring the resemblance of this object to what are commonly identified as 'figurines' in pre-Historic contexts seems to be highly prejudiced."

Regardless of any resemblance between the Berekhat Ram object and an Upper Paleolithic figurine, however, it is inconceivable that the two represent a single cultural artistic tradition, because they are separated in time by over 200 millennia. Even if this object resembles an Upper Paleolithic figurine, it does not follow that it *is* a figurine. Only if it were clearly a representation of a human would this be the case. Neither Goren-Inbar nor Marshack explains the criteria by which they came to this conclusion. This is not to say that they are wrong. It may be that whoever cut the grooves into the pebble intended it to look like a person. However, the contrary is also quite possible.

In my opinion, therefore, the meaning of the Berekhat Ram specimen is equivocal. It is an unusual item for which no "practical" function is apparent. It may have been iconic, as Goren-Inbar and Marshack believe, but it is unique in the context of the Acheulian of the Levant. There is no other artifact like it that might confirm that it carried some public meaning. Thus, even if it was created as representation of a human form, there is no evidence exists that it carried any symbolic (rather than iconic) meaning.

Two objects were recovered from a Mousterian context at Tata, Hungary (Marshack 1976; 1990; Vértes 1959; Vértes 1964) (Figure 5.3). One was a fossil nummulite, approximately 21 mm in diameter. It had a straight crack through the middle and, on one face, a straight scratch at right angles to the crack. The second was a plate of mammoth tooth that, according to Marshack, had been carved into a rough oval. One surface had traces of ochre. The symbolic cultural nature of these objects is unclear. It is difficult to accept a single scratch, even on a fossil, as solid evidence for symbolism. The ochre on the plaque makes it more probable – although in my opinion not certain – that that object had some symbolic meaning. Tata dates to about 100,000 years ago (Schwarcz and Skoflek 1982).

A fragment of bovid rib from Pech de l'Azé II in France bears marks that probably cannot be attributed to use of the bone as a cutting board (Bordes 1969). Nor, judging from the photograph (I have not examined the original), can they be explained as root marks, as Binford (1981:49) claimed. Assuming that the marks were made by a stone tool, the implications are completely unclear. There is little about the marks that indicates a deliberate attempt to make a visually meaningful pattern. Invoking a cultural link to the spaghetti-like markings found on objects from the Upper Paleolithic (Marshack 1976) is meaningless, because the time difference is far too great. At best, the significance of this specimen is equivocal.

There is a perforated wolf canine of uncertain age from the Repolust-Höhle in Austria (the same site where a perforated bone mentioned earlier was discovered) (Bednarik 1992; Mottl 1951). It might be a pendant, or it might have served some other purpose.

The upper Middle Stone Age (MSA) levels of Blombos Cave, Southern Cape, South Africa, produced three items, at least two of which are the kind of thing one would expect to find in the context of elaborated culture. A piece of ochre (Figure 5.4) with one surface artificially flattened bore a pattern of cross-hatched lines "bisected and framed by horizontals" (Henshilwood et al. 2002:279). Each set of parallel lines in the cross-hatching was made at the same time, one set followed by the other, with the horizontal lines made last. A second piece of ochre had a similar

Figure 5.3. Mammoth tooth plate and nummulite from Tata, Hungary. Bar is 1cm. (Drawing after Vértes 1964, plate 5)

although simpler pattern of marks. "The preparation by grinding of the engraved surface, situation of the engraving on this prepared face, engraving technique, and final design are similar for both pieces, indicating a deliberate sequence of choices" (Henshilwood et al. 2002:1279).

The third item was a fragment of bone (probably a mammalian mandible) with a series of subparallel, slightly wavering lines, with one mark crossing the others (D'Errico et al. 2001). Microscopic examination indicated that these marks were made by a stone tool, were broad and flat bottomed, and changed direction slightly with irregularities of the bone surface. Five of them appeared to have been made by repeated passes. These observations indicated that the point, rather than the edge, of one stone tool was used to make all the marks. From this the authors concluded that the marks on the bone constituted deliberate engraving, rather than traces of butchery. Because the piece had almost certainly been broken after being engraved, the implication was that more engraving probably existed on the original, unbroken mandible.

Figure 5.4. Ochre with crosshatched lines from Blombos Cave, South Africa. (Photograph by permission of *Science* and Francesco D'Errico)

There can be little doubt that the marks on the ochre were intended as visual patterns. Although it is possible that these patterns had no referential or ritual symbolic meaning, they are nevertheless exactly the kind of decoration or marking made to convey cultural meaning when socially constructed coding has been elaborated into an overarching cultural system. The fact that a similar pattern appears on two pieces reinforces this conclusion – it would be hard to argue that both represent casual doodling, for example. The interpretation of the marks on the mandible is much less certain, in my opinion, although its association with the two pieces of ochre makes a cultural interpretation more plausible.

Forty-one perforated tick shell (*Nassarius kraussianus*) beads were also recovered from Blombos Cave (Henshilwood et al. 2004) (Figure 5.5). Two of these came from the underlying MSA M2 level, which is provisionally dated to 78,000 years ago. It would be difficult to argue that these are anything other than beads, created for decorative purposes. Ochre residues on some of the beads reinforce this interpretation. Overall, the material from Blombos seems to constitute as solid evidence as one could wish for the elaboration of culture.

The marked ochre, the cut mandible, and two of the beads came from a series of deposits, BBC M1, containing Still Bay points, which date to the Howieson's Poort or before. Five thermoluminescence dates from this series averaged 77,000 ± 6,000 years (Henshilwood et al. 2002:1279). The remainder of the beads came from a level dated by optically stimulated luminescence to 75,600 ± 3,400 years (Henshilwood et al. 2004).

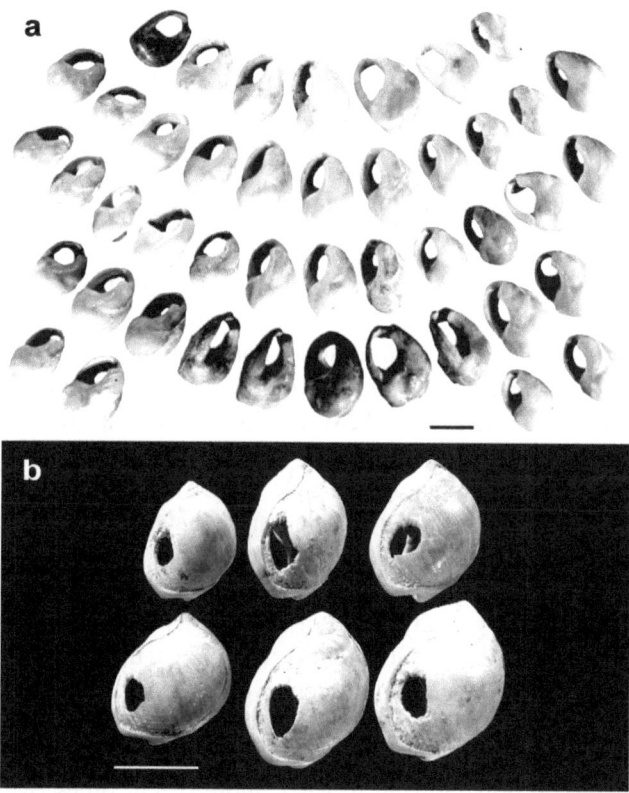

Figure 5.5 Perforated tick shells from Blombos Cave, South Africa. (Photographs by permission of *Science* and Francesco D'Errico)

In the Aterian of Seggédim in eastern Niger, two flakes of quartzitic granite were found that had been perforated by drilling from both sides (Tillet 1978). The flakes were fairly small (3.0 x 1.9 x 0.4 cm and 1.8 x 2.4 x 0.5 cm, respectively). Two other flakes were found with drilling that did not penetrate through them. The meaning of these artifacts is unclear. They may have been intended as beads or pendants, although there is no sign beyond the drilling that they had been shaped for this purpose. The site is undated. Cremaschi et al. (1998) point out that there were wet episodes in the Sahara from about 175,000 to 70,000 years ago, followed by a period of hyperaridity from about 70,000 to 12,000 years ago, with no sign of human occupation. This implies that the Seggédim flakes must be at least 70,000 years old.

Two other possibly symbolic artifacts have been reported from the Aterian. Something that is quite likely a pendant came from Zouhra Cave in Morocco (Debénath 1994:22), and a perforated *Arcularia gibbosula* shell was found at Oued Djebbana, Algeria (Morel 1974:76-77). *Arcu-*

laria gibbosula is a marine species that must have been imported from the Mediterranean, some 200 km away. However, from the photograph it is not clear that the hole was made by humans, and Morel provides no further information in the text. In any case, finite radiocarbon and thermoluminescence dates indicate that the Aterian in the Maghreb persisted beyond 30,000 years ago (Cremaschi et al. 1998; Debénath et al. 1986; Texier et al. 1988; Wrinn and Rink 2003: figure 4). The chronological age of these items is therefore unclear.

After the end of oxygen isotope stage 5, a number of finds constitute at least equivocal evidence for the elaboration of culture.

Léon Henri-Martin reported finding an upper canine of a fox with a hole penetrating into the root at the site of La Quina, Charente, France (Martin 1907-1910:135-138, plate 128). He concluded that this hole represented a failed attempt to make the tooth into a pendant, a not unreasonable conclusion. Unfortunately, he gave no provenience or stratigraphy. Because there were both Aurignacian and Châtelperronian occupations in the immediate vicinity (G. Henri-Martin 1961; 1969; 1976), the possibility that the specimen did not belong to the Mousterian cannot be excluded. (Henri-Martin was far ahead of his time as a zooarchaeologist, but not in terms of excavation techniques). Moreover, the Mousterian of La Quina is not old. Thermoluminescence dates from beds 8 and 6a showed that these beds were deposited between about 48,000 and 40,000 years ago (Mercier and Valladas 1998). Pollen indicates that beds 4b through 2a were laid down during warmer, more recent times (Renault-Miskovsky 1998). Thus, there is no reason to assign a stage 5 date to this specimen.

Three objects that are almost certainly beads were recovered from Mumba Rock Shelter in Tanzania. At least one of these dates to about 52,000 years ago (McBrearty and Brooks 2000:522).

Two ostrich eggshell beads, one complete and one unfinished, were found in the OLP member of Boomplas Cave, Southern Cape, South Africa (Deacon 1995:123). Deacon gives an age of 42,000 years for this member but also states that these finds "require independent dating before their provenance can be accepted."

The MSA (beds 6–9) of the Cave of Hearths, Makapansgat, Transvaal, South Africa, produced an impressive array of apparently symbolic cultural objects: a broken circular piece of ostrich eggshell with a hole in the middle, a tapered piece of horn core with a groove around the thicker end that may have been a pendant, three unmodified quartz crystals and two unmodified fluoride crystals, a "ribbed talc spheroid," and fragments of ochre, one of them ground (Mason 1988: 321-322 and figure 89). A radiocarbon date of ~15,000 years ago was obtained for the MSA levels, but Mason (1988:280) rejected it because of the "inadequate laboratory methods of the time" when the analysis was done. Other than this, there was no direct dating for the MSA at Cave of Hearths.

Mason correlated it typologically with the Middle Pietersburg industry of bed 2 of Olieboompoort, dated by carbon 14 to more than 33,000 years ago (Mason 1988:336). The material from these beds at Cave of Hearths is somewhat unusual for the MSA. Not only were there 12 quite elaborate grindstones (eight upper, four lower) (Mason 1988:320-321, fig. 389), but the industry had resemblances to those of the Late Stone Age (LSA). "The flake typology as a whole verges on the microlithic. A great many flakes and segments are indistinguishable from those in Late Stone Age assemblages. ... The general shape of the Beds 6–9 series suggest the product of a group of Stone Age hunters experimenting with techniques that were to change the Middle Stone Age to the Late Stone Age" (Mason 1988:333). McBrearty and Brooks (2000:500) wrote that "the Later Pietersburg industry (e.g. Cave of Hearths Bed 9) contains backed crescents, and may be attributed to the LSA or the Howieson's Poort." At this point, the dating of these levels must be considered less than solid.

Apollo Cave, Namibia, produced a series of slabs bearing unmistakable representational paintings. McBrearty and Brooks (2000:526) suggested that these might be as old as 59,000 years, on the basis of an isoleucine epimerization date reported by Miller et. al. (1999). However, this date came from the mouth of the cave, where Miller et al. suggest the level may have been truncated. Radiocarbon dates from the top of level E, where these plaques were found, gave a much younger age ($26,300 \pm 400$, $26,700 \pm 650$, and $28,400 \pm 450$) (Wendt 1974:18; 1976:6).

Finally, excavators at La Roche-Cotard, France, discovered a piece of flint with a natural perforation in a Mousterian context (Marquet and Lorblanchet 2003). In this perforation, extending from both ends, was a piece of bone. To the excavators, this resembled a mask. To me, judging from their photographs, this assessment seems highly questionable. Clearly, interpretation in this case is subjective. I am not convinced that the bone did not get into the perforation naturally. However, even if Marquet and Lorblanchet's interpretation is correct, this would simply be a crude and entirely unique iconic object. There is no evidence that it was made to fulfill any cultural role or that it had any cultural meaning.

5.2.4. Ritual

Direct archaeological evidence for ritual (other than that associated with human burials) is naturally much more elusive than is evidence for symbolic meaning in artifacts. Yet a few claims have been made for ritual before the Upper Paleolithic. Probably the best known is that of the so-called cave bear cult (Breuil and Lantier 1959; Maringer 1960), based on what appeared to be intentional burials or positionings of cave bear skulls or bones, notably at Drachenloch, Switzerland (Bächler 1921; 1923; 1940), Régourdou (Bonifay 1962; Bonifay and Vandermeersch 1962) and Les Furtins, France (Leroi-Gourhan 1947), and Salzofenhöhle (Ehrenberg 1953) and Petershöhle, Germany (Hörmann 1933).

However, there have been a series of devastating critiques of this evidence (Jéquier 1975; Koby 1951; Leroi-Gourhan 1964b;   see also Kurtén 1976). There are two grounds for rejecting the claims. First, the most spectacular ones are insufficiently documented. Descriptions and drawings that represent interpretations rather than direct recording of what was found leave one with only two choices: either accept or reject the subjective interpretations of the excavators. There is no way to check their interpretations against the raw data themselves, because the raw data were never recorded.

This is particularly true of Bächler's claims for Drachenloch (see Jéquier 1975). For example, Bächler drew a cist containing cave bear skulls in various different ways in different publications – the size of the cist changed, the orientation of the skulls within it changed, and the construction of the cist changed. Clearly, what he was publishing was not a measured drawing of what was in the ground, but his interpretation. Just as clearly, his interpretation was both loose and subjective. Whatever may actually have been in the site, we cannot accept his reports as convincing evidence. Second, virtually all the reported evidence can be explained taphonomically in terms of rock fall, the natural movements of bones by cave bears, the uneven preservation of bones on surfaces and in crevices or hollows, and the like.

More recently, Lascu et al. (1996) reported three configurations of cave bear bones in Cold Cave in Transylvania. The bones were discovered (judging from their map of the cave) about a kilometer from the entrance and were covered by a calcite crust dated by uranium\thorium to 75,000–85,000 years ago. There was one skull with a rib lying roughly perpendicular across the anterior end, a skull with a radius lying roughly perpendicular a few centimeters in front of the skull, and four crania lying together, with their noses radiating outward. No sign of human presence appeared in the cave, and no remains of any animals besides bears, except for one lamella of mammoth tooth found on a "terrace" (ledge?) of the cave wall. (The authors took this to be evidence of human presence.) They also provided a posthumous drawing, by M. Angus and A. Pasa, of a skull from the Mousterian of "Bear Cave" in Italy. The drawing shows a bear cranium with two long bones on either side, parallel to its long axis, and surrounded by a rectangular arrangement of stones.

The drawing from Bear Cave, like Bächler's reports, constitutes insufficient documentation. It is an interpretation rather than raw data. We can only accept or reject the impressions of the excavators; we cannot judge them against the data. (The same is true of certain as-yet-unpublished data from other sites cited by Lascu and his colleagues.) The patterns from Cold Cave, by contrast, are well documented, but they can all be explained by cave bears' having moved bones about on cave floors.

Piles or other configurations of rocks or clay pellets have also been cited as evidence for early Paleolithic ritual. Blanc (1957) described in some detail a pattern of clay pellets found on a wall in the interior of the cave of Basua, Italy. These were above "a stalagmitic formation, with rounded surfaces, [that] resembles the back of an animal, suggesting a sphinx without a head" (Blanc 1957:113). Because of a blank area on the wall, Blanc suggested that the pellets were thrown at a person standing above this "sphinx," either in a ritual or a game. Whatever their meaning, there is little reason to attribute them to the Paleolithic. The front of the cave had Neolithic and Roman material. The interior had footprints, soot marks, and carbon, but no industry. Blanc's only reason for dating the site to the Paleolithic was Pales's (1954) belief that the footprints most closely resembled those of New Caledonians and that Neanderthal feet had certain analogies to those of New Caledonians. The Basua evidence must be discounted.

Blanc also described a pile of loess pellets found in the site of Achenheim in Alsace, France (Blanc 1957:115-116). He cited a detailed examination of these pellets by Zeuner, who concluded that they were made by humans, probably by rolling balls of loess in their hands. However, there is no evidence that they had any ritual use or symbolic meaning.

In a Mousterian context at El-Guettar, Tunisia, a pile of rounded stones, fragmented bones, and flint was found associated with an artesian spring deposit (Gruet 1959). Neither the bones nor the flint artifacts differed from material in the surrounding sediments, although Gruet did state that several fine Mousterian points were jutting from the upper half of the pile. Gruet cited modern ethnographic ritual uses of rounded stones and also argued that because the pile was apparently at least partly in the water, it probably represented a ritual deposit of some sort. Although this is of course possible, other explanations cannot be excluded. For example, water action had polished the stones on the outer surfaces of the pile, which might indicate that the "cairn" was simply a deposit left behind by the erosion of surrounding sediments.

There is also a collection of fallow deer (*Dama mesopotamica*) bones from level 3 of Nahr Ibrahim, Israel, that Solecki (1982) interpreted as a ritual burial. His argument was based in part on the fact that some of the bones were still articulated and that the sediment was softer in the vicinity of the bones than elsewhere. The strongest evidence was the presence small amounts of very fine fragments of what was probably ochre. Although this might be meaningful, the source of the ochre was apparently immediately adjacent to the site, so the possibility that it was simply brought in accidentally on people's feet cannot be eliminated. In any case, Solecki estimated the age of the site as slightly over 60,000 years, which would postdate stage 5.

5.2.5. Ochre

There can be little doubt that the use of ochre is much older than the Upper Pleistocene. The term *ochre* has been used loosely, and in many cases materials reported as ochre have not been adequately analyzed (Wreschner 1975). Some reports of ochre in early contexts have turned out to be erroneous. Ochre from Upper Bed II at Olduvai (Leakey 1958:1100) proved to be reddened volcanic tuff (Oakley 1981:206-207). What appeared to be flaked ochre at Ambrona, Spain (Howell 1966), was actually naturally fractured siltstone (Butzer 1980). Wreschner (1985:390-392) has expressed considerable doubt about human process-ing of ochre at Terra Amata (de Lumley 1966).

Nevertheless, there is plentiful evidence of the use of ochre or simi-lar materials before the end of the Middle Pleistocene. Fragmentary ochre was found in the Karpathian Formation at Baringo, Kenya, dated to more than 240,000 years ago (McBrearty and Brooks 2000:528). Hematite and abraded hematite were recovered from the MSA of Twin Rivers, Zambia, underlying a speliothem dated to 230,000 +35,000/-28,000 years ago (Barham and Smart 1996). One abraded hematite frag-ment was found associated with a Fauresmith industry at Nooitgedacht 2, Northern Cape Province (Beaumont 1990b:4-5, fig. 4). Soft red hematite was also found with a Fauresmith industry in the Northern Cape Prov-ince at Kathu Pan 1 (Beaumont 1990a:79-80).

A striated piece of ochre and ochre powder were found in the Acheulian site of Beçov in the Czech Republic, along with a quartzite "rubbing stone" (Fridrich 1975; Marshack 1989:12), although it has not been demonstrated that the quartzite stone was used to grind ochre (Wreschner 1985:392). A worn fragment of red ochre was found in the Lower Paleolithic of Achenheim, Alsace (Thévenin 1976), and ochre was found in the Lower Paleolithic of Maastricht-Belvédère in the Neth-erlands (Knight et al. 1995). A single lump of ochre, without signs of use, was recovered at Kabwe (Broken Hill), Zimbabwe (Clark et al. 1947; Leakey 1958). Although Clark et al. believe it was allocthonous, the fact that the site was mined rather than excavated by archaeologists renders the significance of this find dubious at best.

There is also evidence from oxygen isotope stage 5. Ochre, with clear signs of use, was found in all levels of the Klasies River Mouth site. Other rocks that could have been used as pigments were found, but with no signs of utilization (Singer and Wymer 1982:117). Small fragments of low-grade hematite were associated with an early MSA industry at Biesiesput, Northern Cape Province. Soft red ochre was recovered from the late Howieson's Poort and late "'MSA1'" of Kathu Pan 6, Northern Cape Province (Beaumont 1990a:90-91). There is ochre at Porc Epic Cave in Ethiopia that may date from late stage 5. The MSA layers there are dated by obsidian hydration to between 60,000 and 77,000 years ago (Clark et al. 1984:63, 68). It is possible that ochre was being used near the end of stage 5 at Pomongwe Cave, Zimbabwe (Cooke 1963; Walker 1987).

Beaumont (1973:140-141) reported mining of hematite over a long time period from a cliff at Lion Cavern in Swaziland. He believed that at least 1,200 metric tons of pigment were mined there (although he did not explain the basis of this estimate). Even if only a fraction of this was mined during the MSA, it would represent a huge amount of pigment relative to the amount recovered in archaeological sites. The rubble from mining was deposited at the base of the cliff. The lowest level of this rubble included abundant MSA artifacts. Charcoal from the MSA gave carbon 14 dates of 7690 ± 80 B.C., 20,330 ± 1827 B.C., 26,180 ± 260 B.C., and 41,250 ± 1200/1350 B.C., with no infinite dates. Beaumont argued that these dates indicated mixing from younger levels and suggested that because of the MSA industry the level could date back as far as 120,000 years ago. However, it seems unlikely that such an old level would have produced no infinite dates. It makes more sense to consider that the level is not much older than about 42,000 years.

There is, then, no question that ochre was being used long before the end of oxygen isotope stage 5.[*] There is no direct evidence, however, of what it was being used for. In more recent times (notably but by no means exclusively in the Upper Paleolithic of Europe), ochre was used as a pigment. It is therefore logical to hypothesize that it was being used in earlier times to decorate human skin, clothing, rock walls, or other materials. Since all of these are perishable, including rock surfaces exposed to the elements, such decorations could not be expected to survive.

Although this hypothesis is plausible, it is premature, I believe, to conclude that all ochre, most ochre, or perhaps even any ochre was used in this way before about 77,000 years ago. If the earlier MSA of southern Africa was really characterized by fully elaborated culture, then one must

---

[*] Starting in stage 3, the use of manganese became fairly common in Europe (D'Errico and Soressi 2002, 2004). However, this is long after the end of stage 5.

wonder why decoration is not more common on nonperishable materials such as bone, "portable" stones, and shell, as is the case in the LSA.

Certain researchers have proposed alternative uses for ochre, uses that were practical in a material rather than a symbolic sense. One suggestion is that ochre was used to tan hides, but Philibert (1994) refutes most hypotheses concerning how ochre might have been useful in this process. However, she considers only tanning in the strict sense of the word. She does not mention Mandl's explanation (1961:196), cited by Keeley (1980:172), that iron ions would inhibit the action of collagenase, an enzyme that destroys collagen, the primary material in hides. If Keeley's suggestion were true, it would indicate that the use of ochre was a (probably much less effective) forerunner of tanning. Audouin and Plisson (1982) did find experimentally that using ochre on hide had a preservative and drying effect (the latter was more rapid with red than with yellow ochre).

Red and yellow ochre were used in the Upper Paleolithic of France as part of resin-based mastic for hafting stone tools or stone tool components (Allain 1979; Allain et al. 1985; Allain and Descouts 1957). Mixing powders such as clay or ochre with resin and wax is "well known in the ethnographic record [Keeley 1982]; the addition of binding agents minimizes the fragmentation of the resin through crystallization" (Roberts et al. 1998:189; see also Allain et al. 1985:46). Wadley, Williamson, and Lombard (2005; 2004) found evidence for the regular use of ochre in mastic used to haft tools in the post-Howieson's Poort MSA of Sibidu Cave, KwaZulu-Natal.

Audouin and Plisson (1982:55-57) listed ethnographically known uses of ochre: for protection of the hair or skin from vermin and mosquitoes, for protection of the skin from the sun and wind, on the skin for warmth, on wounds (see also Velo 1984; Velo 1986), to neutralize foul odors (see also Cordwell 1985), and to waterproof and protect wood.

In short, there are uses for ochre other than for decoration or art. This is not to say that we know what ochre was used for – there may well have been uses other than those known from the ethnographic record. Nor is it to say that none of the archaeologically known ochre was used for artistic purposes.

Knight et al (1995:87-88) gave examples of markings on ochre that they believed could not be explained in terms of production of powder. These included notches along the edge of a piece of ochre from Hollow Rock Cave, drilling of a piece of ochre from the Klasies River Mouth, and gouging on the surfaces of ochre lumps from Klasies. I have not seen any of these specimens myself, but from the photographs, it appears that however the notches, holes, and gouges were made, powdered ochre was probably produced in the process.

The strongest positive evidence for the decorative use of ochre is a tendency to prefer ochres of a saturated color over duller ones during the MSA (Henshilwood et al. 2001:432-433; Watts 2002). Watts's analysis generally lumps earlier with later MSA data in a way that makes them difficult to interpret for our purposes here. At Blombos, South Africa, the tendency to use saturated colors was evident in the earlier levels, although less saturated colors were also used frequently. There are several possible explanations. Saturated colors might have been preferred for their artistic merits, or they might be indicators of other, more practically useful properties of the ochre. (For example, I mentioned that Audouin and Plisson found that red ochre dried hides more rapidly than yellow.) Some ochre might have been used as pigment, and some in other ways.

Without the final product, we cannot be sure exactly what was being done with ochre during the Middle Stone Age and in what proportion. We certainly cannot assume that it was used in only one way. Producing evidence that ochre was not used in one particular way (e.g., hide processing) would not demonstrate that ochre was used as a pigment. Demonstrating that some ochre was used as a pigment would not demonstrate that all ochre was used as a pigment. In short, there is still considerable uncertainty about the function of ochre in the Middle Pleistocene and early Upper Pleistocene. What we do know is that by 77,000 years ago, at Blombos, ochre was being decorated in a manner that strongly suggests the elaboration of culture.

5.3. IMPLICATIONS OF THE ARCHAEOLOGICAL RECORD

At this point, it would probably be useful to reiterate what I emphasized at the beginning of the chapter, that my only purpose in reviewing the archaeological data was to test the three hypotheses proposed in chapter 3 to explain the elaboration of culture. For this reason, I have not considered data younger than oxygen isotope stage 5, and I have not considered evidence related to burials or to mortuary practices. These are relevant to other questions that are currently hotly debated among Paleolithic archaeologists and human paleontologists, but they are irrelevant to the matter at hand.

Before stage 4, the archaeological record consists of a probably noncultural artifact from Olduvai, rock markings from Auditorium Cave, an object from Tan-Tan that probably had no symbolic or cultural meaning, although it might well have been perceived as iconic, an intriguing artifact from Berekhat Ram whose meaning is unclear, and two enigmatic objects from Tata. In addition, I have mentioned objects that do not meet my minimum criteria for evidence, notably some tools made from fossils or fossil-bearing rocks, and marks on a core from Qafzeh that might indicate its use as a cutting board. Then, from about 77,000 years ago, there are two unquestionably decorated pieces of ochre from Blombos, along with two beads. A level about 1,400 years younger yielded another 39 unquestionable beads. In addition to these objects, ochre was used very sporadically until about 110,000 years ago, after which it was used in larger quantities in at least part of southern Africa. How this ochre was used can be debated.

Let us assume, for the sake of argument, that all the objects I listed were, in fact, symbolic, and that at least some of the ochre was used for decorative or symbolic purposes. The archaeological record is still not what one would expect of a world in which socially created coding had been immediately elaborated into all-encompassing cultural systems a minimum of 127,000 years ago. The evidence is still too sporadic and too narrowly distributed. We would expect a record much closer to what one finds for the Upper Paleolithic or the LSA, where clear artifactual evidence of the presence of elaborated culture, though not universal, is both common and widespread.

What we have before the end of stage 5 is an archaeological record in which the evidence is far from common and, when it exists, is usually equivocal. The major exception is the evidence from Blombos, about 77,000 years ago. There the record is exactly what we would expect from a people with elaborated culture. However, even this is an isolated occurrence, and it dates to some 50,000 years after the end of the Middle Pleistocene, when the data suggest that socially constructed coding was already an integral part of the hominin way of life.

These data, then, are inconsistent with an early, explosive elaboration of culture. They imply, on the contrary, that this elaboration appeared sporadically, increasing very gradually through time. This is entirely consistent with the group benefit hypothesis, but it is inconsistent with the other two hypotheses. If the elaboration of culture was simply a by-product of the use of socially created coding, then it should have happened when such coding was first used, not 50,000 years later. If socially constructed coding was elaborated into all-encompassing cultural systems in order to allay people's fears, sorrows, and the like, then this should have happened as soon as socially constructed coding was available, since all hominins would have been subject to emotional stress. If, however, elaborated systems of culture are a means of giving groups control over the behavior of individuals, then such systems would have arisen only when specific local circumstances made this a more viable way of life than the more individualistic adaptation of all other primate species.

In an earlier publication (Chase 1999), I suggested one set of circumstances under which this could have happened. When environmental conditions lead people to rely heavily on ungulate herd species whose density is low, whose distribution is clumped, and whose location is unpredictable, the best way of exploiting the environment is to maintain widespread networks of people who share information. In widespread networks, however, cheating is a major problem. Even if cooperation would benefit everyone, it would not necessarily be in the best interests of an individual to cooperate. One way of overcoming the natural tendency to cheat would be to use cultural ideology, backed up by the emotional force of ritual.

I cannot pursue this hypothesis further in this book, for the simple reason that at the present time the archaeological record is insufficiently fine-grained to test it. I chose instead to propose the three general hypotheses described in chapter 3. In a certain sense, these could be considered families of hypotheses. For example, my more specific hypothesis is simply one of a series of possible group benefit hypotheses. However, the more general hypotheses have the advantage of being testable with the archaeological record as it exists today.

Parenthetically, I should note that I have not considered another hypothesis because I consider it untestable. Knight, Powers, and Watts (1995) have suggested that the origin of culture is rooted in women's use of ochre to simulate menstruation, with the purpose of inducing males to bond with females and to help them provision their young. The hypothesis posits not just the existence of symbols but specific symbolic meanings. Given that what we have in early archaeological sites with ochre is not even symbolic representations but at best the raw material from which symbols might have been made, it seems to me that hypotheses about what these invisible symbols meant are essentially untestable. (I must add that I find it unlikely that even 110,000 years ago males would have been unable to tell the difference between ochre and menstrual blood, at least after the first few months.)

There remains one last implication of these findings. It is now generally accepted that Neanderthals were responsible for the Upper Paleolithic Châtelperronian industry of France (Hublin et al. 1996; Leroi-Gourhan 1959; Lévêque and Vandermeersch 1980). Châtelperronian levels at Arcy-sur-Cure and other sites have produced a series of artifacts that, like the material from Blombos, are what one would expect to find in the context of elaborated culture (Leroi-Gourhan and Leroi-Gourhan 1965; Taborin 1988; Taborin 1990; White 1993). Grooves have been cut around the tips of carnivore and bovine teeth, reindeer accessory toe bones, and a shell, apparently for suspension. There are animal teeth and bones that have been pierced, again apparently for suspension. There are partial worked ivory rings. There are also a fragment of bird bone from Arcy-sur-Cure and a bone awl from the Grotte des Fées at Châtelperron with thin cut marks that appear to have been intended as decoration (Debénath et al. 2002; Leroi-Gourhan and Leroi-Gourhan 1965).

Any one of these could be dismissed as not having been symbolic or cultural, but they closely resemble pendants, rings, and decorated objects from the Aurignacian, both in appearance and technique of manufacture (White 1993; 2001; 2002). Especially since there is a repeated pattern to these objects, it seems that the best explanation is that they had some cultural meaning.

Yet such material is absent from the great bulk of the archaeological record attributable to Neanderthals. Moreover, DNA extracted from fossil remains is providing direct evidence that Neanderthals may have been a separate species from our own (Beerli and Edwards 2002; Caramelli et al. 2003; Gutiérrez et al. 2002; Krings et al. 1999; Krings et al. 1997; cf. Nordborg 1998; Relethford 2001; Scholz et al. 2000; Tschentscher et al. 2000). There are really only four ways of resolving this dilemma:

1. The Arcy material may have been created by *Homo sapiens* instead of *H. neanderthalensis*.
2. Neanderthals and moderns may both belong to the same species, *H. sapiens*.
3. The cognitive capacity for elaborated culture may have evolved before the split between *H. sapiens* and *H. neanderthalensis*.
4. These apparently cultural artifacts had nothing to do with elaborated culture.

Note that possibilities 2 and 3 imply that the cognitive abilities for cultural elaboration evolved long before the Châtelperronian. In these scenarios, Neanderthals had the symbolic and cultural abilities that underlie the elaboration of culture. They would not have made such artifacts unless they understood and were motivated by culture. Even if they imitated or emulated *Homo sapiens,* they did so because they, too, were capable of becoming fully cultural beings.

If either alternative 2 or 3 is true, then the long absence of evidence for elaboration in the archaeological record of the Neanderthals indicates that cultural elaboration occurred not when it was genetically possible but when local circumstances made it advantageous. This is one more thread of evidence in favor of the group benefit hypothesis. It does not matter whether the Châtelperronian material was an "acculturation" from the Aurignacian or not (D'Errico, Zilhão et al. 1998; Mellars 1999; 2000b; 2000a; 2005; White 2001; Zilhão 2000; 2001; Zilhão and D'Errico 2000). Nor does it matter if Arcy-sur-Cure was an isolated instance and that there never was a fully elaborated Neanderthal culture, as long as Neanderthal culture was becoming elaborated.

If neither alternative 1, 2, nor 3 is true, then 4 must be true. Wynn and Coolidge (Coolidge and Wynn 2005; Coolidge and Wynn 2004; Wynn and Coolidge 2004) argue that Neanderthals differed neurologically from modern *Homo sapiens* in that they had a more limited working memory. They believe this accounts for the lack of innovation throughout most of the Neanderthal archaeological record. If this is true, it might imply that Neanderthals did not possess the cognitive abilities needed for the elaboration of culture and that neither 2 nor 3 is possible. However, there are possible alternative inferences.

First, it might be that this lack of innovation is attributable not to a lack of working memory but to a lack of elaborated culture. As long as artifacts are produced with purely practical ends in mind, they will lack most of the temporal and spatial variability and all of the symbolic content found in those produced with cultural concerns in mind. In this case, either 2 or 3 might well be true.

Second, even if Neanderthals did possess more limited working memory, it might still be that 3 is true – that Neanderthals possessed sufficient cognitive ability for elaborated culture. Coolidge and Wynn suggested that Neanderthals learned to make Châtelperronian artifacts by emulation from the Aurignacian. That is, they did not learn the methods and motor habits of the Aurignacian, but rather they recognized the goals represented by the artifacts and found their own ways of achieving those goals. If this is true, then the Arcy data imply that late Neanderthals in southwestern Europe were elaborating their culture for some reason. Even if they differed from their *Homo sapiens* contemporaries in terms of working memory, they too possessed the neural prerequisites for elaborated culture.

6

CONCLUSION

In the preceding pages, I have confronted readers with a large number of ideas and arguments. There is a real possibility that I have hidden the forest in an overabundance of trees. It will be useful, therefore, to review what I have written, not to summarize it but to emphasize what I consider to be most important. It will also be useful to review the state of our knowledge or lack thereof. (Readers who have come directly to this chapter looking for a summary of the book should turn instead to chapter 1.)

My primary purpose has been to focus the attention of Paleolithic archaeologists on the origins and evolution of human culture. Most North American Paleolithic archaeologists were trained in departments of anthropology, yet we have concentrated our research almost entirely either on intelligence and cognition or on language and symbolism (the latter two being only a subset of culture). Culture, however, is of enormous importance to the human way of life, and any account of hominin evolution that ignores it is woefully incomplete.

My main thesis is that human culture is a manner of governing behavior, one that coexists with ways of doing so that we share with other species but that is unique by virtue of its emergent nature. Our behavior, like that of all mammals, is guided by noncultural coding that is in part genetically determined, in part individually learned, and in part socially learned. Such noncultural coding, even the part of it that is learned socially, can be understood at the level of the individual. Cultural coding, because it is *created* socially, cannot be understood at the level of the individual. This emergent property defines what I call culture and is its most essential aspect.

The elaboration of culture into all-encompassing systems and our willingness to let cultural coding motivate our behavior are secondary, because they depend on the existence of socially created coding, and because it is possible that socially created coding existed without them for a

significant period of time. Nevertheless, they are ubiquitous among ex-
tant human societies, and they are of central importance to the way all
Homo sapiens live today. Our willingness to let coding that is socially
created motivate our behavior changes the relationship between the individual and the social group. Because culture has been elaborated into all-
encompassing systems that include almost everything we perceive, think,
or do, this new relationship colors all of human life. I consider these
ideas, which I presented in detail in chapter 2, to be the core of this book
and the basis of everything discussed in subsequent chapters.

There are two qualifying considerations I should make clear. First, it
was not my intention to present a "correct" definition of culture, one that
I believe should be adopted by all anthropologists. I am not suggesting
that other definitions of culture are "wrong." In fact, many anthropological theories of culture are compatible with what I describe. Even memetics, which I criticized in chapter 2, I find inadequate not because it is
wrong but because it is incomplete.

Second, when I began to research the primatological literature, I did
not expect culture to be entirely confined to humans. I expected to find at
least partial or rudimentary forms of emergent, socially created coding in
other species, if only among our nearest relatives, much as we find rudimentary symbolic abilities among the great apes. However, as far as I
can determine, this is not the case. Although other species seem to have
the mental abilities needed to create coding socially, they apparently do
not do so. On empirical rather than a priori grounds, therefore, I concluded that among living species, humans are unique in this respect. Like
all empirical judgments, this one may be nullified by new research or by
more thorough analysis by primatologists.

Because I am a Paleolithic archaeologist, I could not stop at describ-
ing culture. I had to ask when, why, and how such a fundamentally important part of the human way of life came into being. I tried to answer
this question in chapters 3 through 5. How well I have succeeded, and
the problems I have encountered, are worth looking at.

I did not propose hypotheses to explain all three aspects of culture. I
did not consider hypotheses to explain the origins of simple socially created coding. The benefits of such coding are so obvious in terms of increasing the effectiveness of cooperative behavior (where evolutionary
altruism is not involved) that there is no theoretical problem explaining
its origins. Indeed, it might be more productive to ask why other species
have never developed this way of coordinating their behavior, especially
the apes, who seem to have the necessary social and cognitive abilities.

I did propose three very general hypotheses to explain the elaboration of cultural coding into all-encompassing systems:

1. That this was simply a by-product of socially created coding and of the cognitive and social capacities underlying it (the by-product hypothesis)
2. That as soon as the ability to create codes socially became available, it was elaborated into cultural beliefs and practices that could allay the emotional stresses inherent in the lives of all mammals (the anxiety hypothesis)
3. That the elaboration of cultural coding provided social groups with a means of influencing the behavior of individuals for the benefit of the group, even at the expense of the individual's evolutionary fitness (the group benefit hypothesis)

I discussed the ways in which these three hypotheses could be tested against one another. Some tests depended on data from disciplines such as psychology that deal with living humans. These I could not address.

However, I found one test based on archaeological and paleontological data that could distinguish between the by-product and anxiety hypotheses, on the one hand, and the group benefit hypothesis, on the other. It follows from the first two hypotheses that the elaboration of culture was, on a species-wide scale, simultaneous with the origins of socially created coding. The group benefit hypothesis permits an indefinite time lag between these two events.

Because cultural motivation of individual behavior may be archaeologically invisible, I did not attempt to test any hypotheses for its origin. This will have to be left to other disciplines. However, the group benefit hypothesis includes cultural motivation of individual behavior as a sine qua non. To the extent that the archaeological record supports the group benefit hypothesis, it implies that cultural motivation evolved either before or simultaneously with the elaboration of culture. I discussed mechanisms by which group-beneficial predispositions, such as a willingness to let culture motivate one's behavior, could have evolved.

I reached three conclusions based on the available primatological, paleontological, and archaeological data.

- Living primate species other than humans do not create coding through social interaction. This means that the origin of what I call culture lies somewhere on the hominin line.
- The data provided by fossil endocrania, by the reconstruction of fossil vocal tracts, and by zooarchaeological evidence for cooperative hunting all suggest (albeit tentatively) that socially constructed coding (including referential language) was a part of hominin adaptation by the end of the Middle Pleistocene, and perhaps long before.

- The archaeological record does not provide evidence of widespread elaboration of culture for at least 50,000 years after the Middle Pleistocene.

The by-product and anxiety hypotheses must be therefore be rejected in favor of the group benefit hypothesis.

However, it is worth examining the basis for these conclusions. Doing so paints a rather stark picture of the state of our knowledge today. There are several weaknesses in the chains of argument and data on which the conclusions are based.

It is possible to challenge many of the bridging arguments involved. For example, many arguments linking language with endocrania or with vocal tract reconstructions are open to question. Likewise, many of my arguments in chapter 3 are based on research in disciplines outside archaeology, research that is still in progress. Understanding individual versus group benefits of culturally motivated behavior will necessarily depend on developments in evolutionary theory as it applies to individual versus group versus multilevel selection.

All too often researchers have not tested alternative hypotheses. This is especially true of arguments linking endocast morphology or tool making to language. There may be data to support particular hypotheses regarding the origins of language, but until they are tested against alternative hypotheses, we cannot accept them with a high degree of confidence. Testing a single hypothesis tells us only whether that explanation is possible. Testing multiple hypotheses tells us which of them best fits the data. Failing to test one plausible hypothesis against another plausible hypothesis leaves us with no way of choosing between them.

In many cases, data needed to test a hypothesis are unavailable. For example, we do not know the relationship between basicranial flexion and the production of formant frequencies, beyond the fact that they are correlated in child development. We have no objective measures of the probability that using a bone as a cutting board would produce a given pattern of cut marks. Many such deficits can be remedied by further research.

Data needed to test hypotheses may be unavailable simply because the relevant questions have never been asked. For example, a great deal of research is being done on whether or how humans evolved a tendency to altruism – a predisposition to help others at the expense of one's own individual fitness. I suggested in chapter 3 that humans might instead have evolved a susceptibility to culture, a propensity not to help others but to obey the dictates of socially created coding, even at the expense of their individual fitness. In many cases, this would lead to the same behavior – helping others – but the underlying psychological mechanism would be different. Because, as far as I know, this has never been suggested before in this form, I have found no research on the subject and no data that might bear on the hypotheses I have proposed.

New data are constantly coming to light, and new analyses of old material bring about new interpretations. At the present time, this seems to be especially true of evidence for the use of ochre in the Upper Pleistocene (Soressi and Henshilwood 2004; Thompson et al. 2004). Moreover, given the current state of our discipline, every Paleolithic archaeologist has a different set of criteria for evaluating the data and will interpret them differently.

Finally, understanding the origins of culture in all its aspects cannot be accomplished by archaeologists alone. Researchers from multiple disciplines will have to address the question, and research in each field will have to be coordinated and made relevant to what is being done in others.

In summary, this book is only a first step. It is common practice in academia to introduce a new theoretical argument along with enough supporting data to present the reader with a fait accompli. To conform to this model, I would have had to present a thoroughly tested, solidly confirmed account of the origins of human culture. This I have been unable to do.

Even if I were under the illusion that I had done so, I would be wrong. I am sure that some readers will disagree with my evaluation of the archaeological, fossil, and primatological data. Others undoubtedly will be unsatisfied with the hypotheses or with the test implications I presented in chapter 3. It was not my purpose, however, to present the final word on the subject. This essay is a starting point, and I will consider it successful if either of the following comes to pass.

First, I hope that my discussion of culture will stimulate interest in the way humans create the cultural coding that governs our behavior. I hope it will stimulate examination of the way our willingness to let socially created coding motivate our actions affects the relationship between the individual and the group, in terms of both behavior and natural selection.

Second, I hope that this effort will focus the attention of Paleolithic archaeologists on the evolution of culture as a phenomenon. It is not necessary that archaeologists accept my definition of human culture. However, it *is* necessary that Paleolithic archaeology address the issue of culture. If culture is more important to human adaptation than it is to the adaptation of any other species, then we cannot pretend to have understood human evolution until we have also understood the origins of human culture.

A

APPENDIX AND GLOSSARY

A.1. USING THE APPENDIX

This appendix is intended as a glossary for readers who are not specialists in archaeology and human paleontology. However, although some terms can be briefly defined in the manner that is common in glossaries, others are better presented in the context of an overview of a topic. Terms of the latter sort appear in the glossary, but the reader is referred to the appropriate overview. There are overviews of chronology, taxonomy, and of archaeological terminology. Within these overviews, terms appearing in the glossary are in bold font.

For readers who would like a more thorough review of these topics, I suggest Klein 1999.

A.2. GLOSSARY

Acheulean	See overview of archaeological periods and industries.
allometry	Changes in the shape of an organ (or organism) as a function of a change in size. Allometric relationships may be nonlinear, and different parts of the organ may have different allometric relationships to the overall size.
artiodactyl	A member of the order Artiodactyla – generally, a hoofed mammal with an even number of toes.
Aterian	See overview of archaeological periods and industries.

aurochs	Wild cattle (*Bos taurus*), now extinct.
backing	See overview of stone tool technology.
biface	See overview of stone tool technology.
bifacially flaked	See overview of stone tool technology.
bovine	A member of the subfamily Bovinae, which includes cattle (*Bos*) and bison.
canine	In carnivores and most primates, the tooth commonly referred to as a fang. In hominins, the canine teeth have been reduced in length to the general height of the tooth row, and the shape has been modified.
chert	A rock made up of very small quartz crystals. The small size of the grains makes it easy to knap (q.v.), since it fractures smoothly and predictably. Terminologies for such rocks (chert, flint, chalcedony, etc.) are somewhat variable.
core	See overview of stone tool technology.
cortex (re. brain)	See neocortex
cortex (re. stone tools)	The surface of an unbroken nodule of stone. It is usually distinguishable from the surface left by knapping (q.v).
cranium	The skull, exclusive of the mandible, or lower jaw.
Emireh point	A Levallois point (q.v.) with the base thinned.
encephalization	The ratio of brain size to body size, adjusted for allometry.
epiglottis	A leaf-shaped fibrous projection behind the tongue and in front of the upper opening of the larynx. It is attached to the hyoid bone by a ligament.
exaptation	A trait that serves an adaptive purpose other than the one responsible for its origin.
exogamous	(Of a social group.) Requiring that members marry outside the group.
femur	The bone of the thigh.

flake	See overview of stone tool technology.
flint	See "chert."
frontal lobe	The paired lobes of the cerebral surface located at the anterior part of the brain. Separated from the parietal lobes by the fissures of Rolando.
hammer stone	See overview of stone tool technology.
handax	See overview of stone tool technology.
hematite	See "ochre."
Holocene	See overview of chronology.
horn core	The bony core of a horn. The chitenous exterior usually decays long before the archaeologist uncovers the specimen.
Howieson's Poort	See overview of archaeological periods and industries.
hyoid	A U-shaped floating bone in the throat that is connected by muscles and ligaments to the cranium, mandible, tongue, and larynx.
in situ	In place; not removed from its original context in geological or archaeological deposits.
indirect percussion	See overview of stone tool technology.
industry	See overview of archaeological periods and industries.
interglacial	See overview of chronology.
isoleucine epimerization	A dating method based on postmortem changes (recemization) of the amino acid in eggshell.
knapping	See overview of stone tool technology.
larynx	A cartilaginous structure at the top of the windpipe that contains the vocal cords.
Later Stone Age	See overview of archaeological periods and industries.
Levallois point	"A triangular flake with a central ... ridge, sometimes possessing a triangle at the base due to prior removal of a small flake. It is this

	medial ridge which … predetermines the triangular form" (Tixier, 1960 in Debénath and Dibble 1994:50).
Lower Pleistocene	See overview of chronology.
LSA	See overview of archaeological periods and industries.
M1	First molar.
M3	Third molar.
mandible	The lower jaw.
manuport	Something not naturally occurring in an archaeological site that is believed to have been brought to that location by hominins.
microlith	See overview of stone tool technology.
Middle Paleolithic	See overview of archaeological periods and industries.
Middle Pleistocene	See overview of chronology.
Middle Stone Age	See overview of archaeological periods and industries.
moiety	Some societies are organized, in terms of kinship and other social relationships, into complementary halves, or moieties.
morpheme	The smallest unit of sound in a language that has some meaning. For example, the word "cats" consists of two morphemes: "cat," which means a certain animal, and "s," which means more than one.
Mousterian	See overview of archaeological periods and industries.
MSA	See overview of archaeological periods and industries.
neocortex	The gray matter covering the exterior of the cerebrum or forebrain. Associated with the highest levels of information processing.
Neolithic	In rough terms, the archaeological time period between the beginnings of animal and plant domestication and the development of com-

	plex societies, city-states, or the common use of metal. A term that is applied variably from region to region.
nummulite	A large protozoan of the subclass Foraminifera. Nummulites had calcium carbonate shells, which are common as fossils in some ancient (Eocene) limestones.
ochre	A naturally occurring oxide that can be ground or crushed and used as a pigment. Red ochre consists of iron oxide or hematite, and brown and yellow ochres, of other materials.
Oldowan	See overview of archaeological periods and industries.
ontogeny	The growth and development of an organism from a fertilized ovum to an adult, and the changes that occur during this process.
oxygen isotope stage	See overview of chronology.
phalanges	The bones of the finger or toe.
Pleistocene	See overview of chronology.
Pliocene	See overview of chronology.
preadaptation	See "exaptation."
Prodnik	A variable type of bifacially worked tool, usually asymmetrical, often D-shaped, with one straight edge and usually with a special resharpening called "tranchet" at the tip.
radiocarbon	A dating method based on the radioactive decay of naturally existing carbon 14.
retouch	See overview of stone tool technology.
rock shelter / shelter	A sheltered area under overhanging rock. Shallower than a cave.
scapula	Shoulder blade.
secondary burials	Burials of bones after the soft tissue has decayed.
speleothem	A general term that covers stalactites, stalagmites, flowstone, etc., that is, carbonate deposits within a cave.

subglottal   Below the glottis, the middle portion of the larynx in which the vocal cords are located.

tang   See overview of archaeological industries.

taphonomy   Strictly speaking, the "laws of burial" or the processes by which a fossil or artifact enters the geological or archaeological record. More generally, the factors (or the study of those factors) that destroy or distort part of the fossil or archaeological evidence.

technology   See overview of stone tool technology.

temporal lobe   The paired lobes of the cerebral surface located laterally and on the lower parts of the brain. Separated from the parietal lobes by the fissures of Sylvius and from the posterior occipital lobe by the parieto-occiptal fissures.

terminus ante quem   The latest point in time at which an event could have occurred. The event could have occurred at any earlier time.

terminus post quem   The earliest point in time at which an event could have occurred. The event could have occurred at any later time.

thermoluminescence   A dating method based on measuring the radiation absorbed by rocks since the last time they were heated. Useful archaeologically for rocks heated in prehistoric fires, etc.

thoracic vertebrae   Vertebrae of the chest – those to which ribs are articulated.

tuff   Rock made up of small volcanic fragments.

typology   See overview of stone tool technology.

ungulate   Hoofed mammal.

Upper Paleolithic   See overview of archaeological periods and industries.

Upper Pleistocene   See overview of chronology.

A.3. TOPICAL OVERVIEWS

A.3.1. Chronology

In geology, the time since the formation of the Earth has been divided into three eras, which in turn are subdivided into periods and epochs. The time that concerns us here lies within the most recent era, the Cenozoic. The **Pleistocene** epoch spans the time from approximately 1.75 million years ago (when it superseded the **Pliocene**) until about 10,000 years ago (the beginning of the present **Holocene** epoch). The Pleistocene is often referred to popularly as the Ice Age, although the term is somewhat misleading. It was in fact characterized by major fluctuations in climate, some of them somewhat warmer than today.

The most comprehensive record of these fluctuations comes from changes in the ratios of oxygen isotopes (^{16}O and ^{18}O) in sediment cores recovered from the deep sea floor (Figure A.1). Because the ratio fluctuates as a function of temperature, these fluctuations, or **oxygen isotope stages**, serve as the backbone of Pleistocene chronology and are especially useful toward the end of the Pleistocene. However, the beginning of the Pleistocene is defined by faunal changes in Italy that date to somewhere between 2.1 and 1.6 million years ago. The break between the earlier **Lower Pleistocene** and the **Middle Pleistocene** is defined on the basis of a reversal of the Earth's magnetic field about 780,000 years ago. The beginning of the **Upper Pleistocene** is dated to the beginning of oxygen isotope stage 5, about 127,000 years ago.

A good, thorough, but readable summary of the geological chronology and of the dating techniques relevant to hominin evolution can be found in Klein 1999.

Note: Do not confuse Lower, Middle, and Upper Pleistocene, which are geological time spans, with Lower, Middle, and Upper Paleolithic, which are archaeological terms.

A.3.2. Taxonomy

The biological taxonomy of our ancestors and nearest relatives is in a state of flux. Over the past few years, taxa above the genus level have often been merged, while fossils within the genus *Homo* have been divided into a number of different species. There is little agreement, especially in detail, about how living or fossil primates should be classified. However, the following is a fairly typical taxonomy.

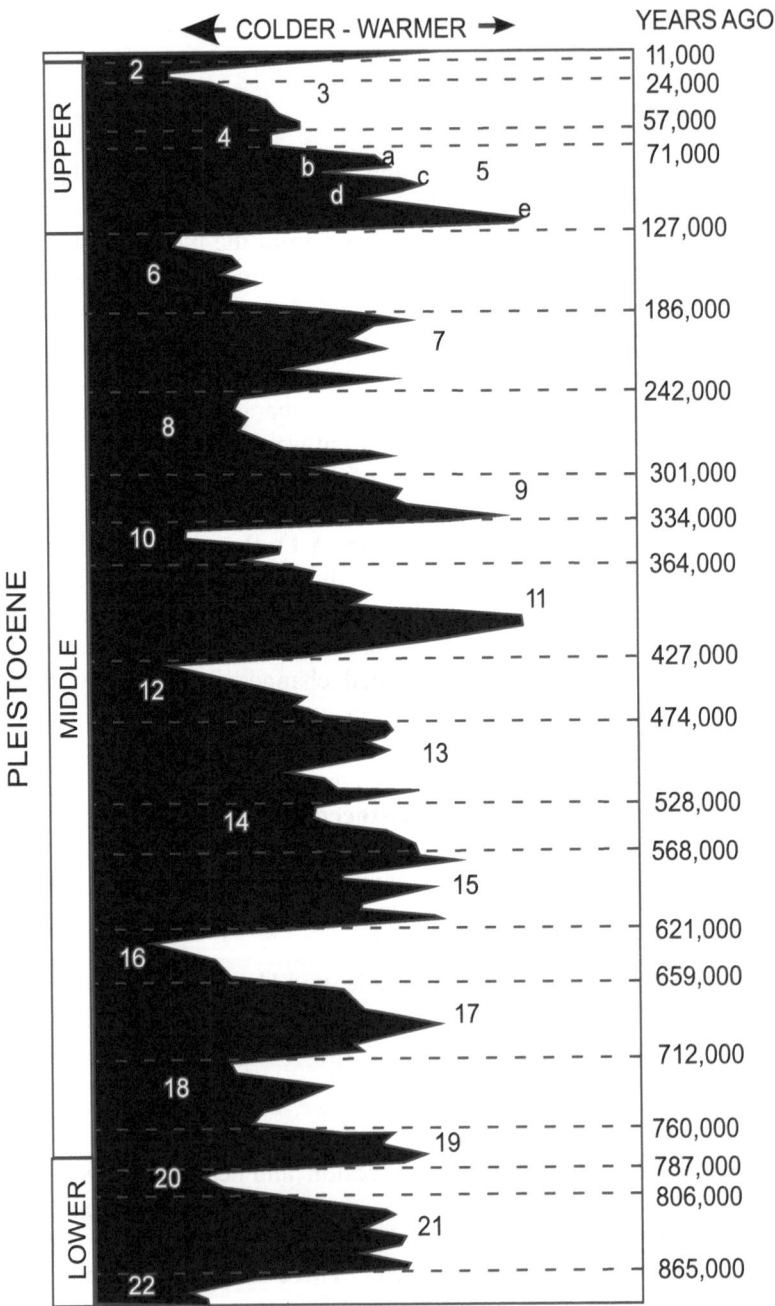

Figure A.1. Oxygen isotope stages, with approximate dates. (Adapted from Klein 1999, figure 2.12.)

Order Primates
    Suborder Anthropoidea (monkeys, apes, and humans)
        Infraorder Catarrhini (Old World monkeys, apes, and humans)
            Superfamily Hominoidea (apes and humans)
                Family Hylobatidae (siamangs and gibbons)
                Family Hominidae (great apes and humans)
                    Subfamily Ponginae
                        Tribe Pongini
                            Genus *Pongo* (**orangutans**)
                    Subfamily Homininae
                        Tribe Panini
                            Genus *Gorilla* (**gorillas**)
                            Genus *Pan*
                                Pan troglodytes (**chimpanzees**)
                                Pan paniscus (**bonobos,** or pygmy chimpanzees)
                        Tribe Hominini (**hominins**)
                            Genus *Australopithecus*
                            Genus *Homo*

The informal term "great apes" refers to orangutans, gorillas, chimpanzees, and bonobos.

The taxonomy of hominins is even more controversial and more unsettled than that of primates in general. The following, therefore, should not be taken as given truth, or even as a summary of the topic, but simply as the minimum that need be said to make it possible for nonspecialists to understand my discussion of culture.

The earliest hominins are generally assigned to the genus *Australopithecus.* This genus was ancestral to our genus, *Homo,* but also contained at least two extinct side branches. These are categorized by some as different species of *Australopithecus* or as two genera, *Australopithecus* and *Paranthropus.*

The earliest members of the genus *Homo* are generally classified into two species, *Homo habilis* and *H. rudolfensis,* with a certain amount of disagreement about the relationships between them and with later hominids.

About 1.9 million years ago, a new species appeared in Africa. Again, the extent of this species and its relationships to later species are controversial. One school of thought now assigns them all to a single species, *H. erectus.* Another would reserve this term for the later, generally Asian populations and assign the ancestral African population to *H. ergaster.*

Hominin populations in Europe that were more archaic than modern humans but less so than *H. ergaster* were once referred to as "archaic *H. sapiens*" but are now often assigned to the species ***H. heidelbergensis.***

There is enormous controversy about the relationship between Neanderthals and modern humans. One school of thought places them in two separate species, ***H. neanderthalensis*** and ***H. sapiens.*** Neanderthals would in this view have evolved from *H. heidelbergensis* and have been replaced, with little or no genetic admixture, by populations of *H. sapiens* migrating from Europe. The other school sees both as subspecies (*H. sapiens neanderthalensis* and *H. sapiens sapiens*) of a species that evolved in situ across much of the old world.

A.3.3. Archaeological Terminology

Paleolithic archaeological assemblages are usually assigned to categories according to the stone tools found in them. Definitions in archaeological taxonomy require some understanding of how stone tools are made.

A.3.3.1. Stone Tool Technology

Because archaeologists use the term "tool" in a specialized way, worked stone artifacts are usually referred to as **lithics.** Almost all the lithics in the periods of interest in this book were made by **knapping**, that is, by removing flakes either from a block or nodule of stone (called a **core**) or from a larger flake.

Flakes may be removed either by percussion or by pressure. For percussion flaking, either hard hammers (stone) or soft hammers (bone, antler, or hard wood) could be used. Stones used as hammers are called **hammer stones**. A specialized form of flaking called **indirect percussion** involves striking a piece of bone or antler held against a core or the edge of a flake. Force transmitted through the bone or antler removes a flake.

The removal of series of small flakes to shape or to resharpen tool edges is called **retouch.** Removal of a series of small flakes that blunts an edge is called **backing.**

Flaking or retouch may be **unifacial** (flakes removed from one surface) or **bifacial** (flakes removed from opposite surfaces.

Tools made by removing flakes from cores are called **core tools.** Tools made by retouching or reshaping flakes are called **flake tools.** Both of these are classified as **tools.** Flakes and other unretouched artifacts are referred to as **debitage.**

The classification of lithic artifacts in terms of their form or morphology is referred to as **typology.** The methods used in producing lithic artifacts are referred to as **technology.**

A.3.3.2. Periods and Industries

An **assemblage** consists of all the lithic and other artifacts coming from a single provenience, usually one geological level or stratum at one site. Lithic assemblages that resemble one another are classified as **industries** (sometimes misleadingly referred to as "cultures"). Industries are the basis of archaeological classification during the Paleolithic. It should be noted, however, that the definition of industries is a rather subjective affair, and that in some cases there is disagreement over details. I describe here only industries that are mentioned in the text.

The **Oldowan** is the oldest industry. It began more than 2 million years ago in Africa and lasted until about 1.7 or 1.6 million years ago. It is characterized by rather unsophisticated knapping techniques. It contains both flake and core tools. Similar industries are found until rather recent times in some places, although on chronological grounds these would never be classified as Oldowan.

Assemblages are assigned to the **Acheulean** (or Acheulian) industry primarily on the presence of **bifaces** or **handaxes,** which are core tools worked bifacially, more often than not over the entire surface. Flaking techniques are often highly sophisticated. Bifaces of various kinds are also accompanied by flake tools. The Acheulean is found throughout Africa and western and much of central Eurasia. It grew out of and replaced the Oldowan in sub-Saharan Africa and persisted until about 250,000 to 200,000 years ago. It should be noted that non-handax industries coexisted with the Acheulean.

Mousterian industries replaced the Acheulean in North Africa and Eurasia. These grew out of the Acheulean and differ from it primarily in the absence of bifaces or handaxes (although these do appear in some Mousterian industries).

In sub-Saharan Africa, the industries that replaced the Acheulean are called **Middle Stone Age** (or **MSA**). Of these, the most remarkable is the **Howieson's Poort,** which dates to about 70,000 to 60,000 years ago. It is characterized by geometric microliths – small, regularly made flakes that were almost certainly hafted, perhaps in series, as parts of compound tools.

All the industries that date to before the end of the Pleistocene are considered **Paleolithic.** In North Africa and Eurasia, the Acheulean and contemporaneous industries are called **Lower Paleolithic.** The Mousterian industries that followed them in North Africa and Eurasia are usually referred to as **Middle Paleolithic.** Most Middle Paleolithic industries are Mousterian in nature. Starting about 40,000 years ago, in Europe, these were replaced by industries characterized by the regular production of blades (long narrow flakes), bone tools, and undisputed art. These are referred to as **Upper Paleolithic,** and the term is applied to similar industries in the rest of Eurasia. In sub-Saharan Africa, the Oldowan and Acheulean are called the **Early Stone Age (ESA)**. The origins of the **Later Stone Age (LSA)**, which replaced the MSA, are poorly known, due to a paucity of sites between 50,000 and 20,000 years old. LSA industries tend to be characterized by high frequencies of microliths.

The **Aterian,** a North African industry, is characterized by tools (points, scrapers, etc.) with an elongated **tang** at the base, presumably for hafting. Chronologically, the Aterian lies at the Middle to Upper Paleolithic boundary, although there are disagreements about the dating.

Note: Archaeological terminology contains several pitfalls for the unwary: (1) Upper, Middle, and Lower Paleolithic should not be confused with Upper, Middle, and Lower Pleistocene, which are geology-based chronological terms. (2) Apparently functional terms such as "handax" and "scraper" are used in a purely morphological sense by archaeologists, without any implications about the actual uses of the tools.

REFERENCES

Adam, K., 1951, Der Waldelefant von Lehringen, eine Jagdbeutes diluvalien Menschen, *Quartär* **5**:75-92.

Adam, K. D., 1985, The chronological and systematic position of the Steinheim skull, in: *Ancestors: the Hard Evidence*, E. Delson, ed. Alan R. Liss, New York, pp. 272-276.

Aiello, L. C. and P. Wheeler, 1995, The expensive-tissue hypothesis: The brain and the digestive system in human and primate evolution, *Current Anthropology* **36**:199-222.

Albrecht, G., C.-S. Holdermann, T. Kerig, J. Lechterbeck and J. Serangeli, 1998, "Flöten" aus Bärenknochen - Die frühesten Musikinstrumente? *Archäologisches Korrespondenzblatt* **28**:1-19.

Albrecht, G., C.-S. Holdermann and J. Serangeli, 2001, Towards an archaeological appraisal of specimen No. 652 from Middle-Paleolithic level D/(layer 8) of the Divje Babe I, *Arheološki Vestnik* **52**:11-16.

Alcock, J., 2001, *The Triumph of Sociobiology*, Oxford University Press, Oxford.

Alexander, R. D., 1979, Evolution and culture, in: *Evolutionary Biology and Human Social Behavior*, W. Irons and N. A. Chagnon, eds., Princeton University Press, Princeton, pp. 59-88.

Alexander, R. D., 1987, *The Biology of Moral Systems*, Aldine de Gruyter, New York.

Alexander, R. D., 1989, Evolution of the human psyche, in: *The Human Revolution: Behavioural and Biological Perspectives on the Origins of Modern Humans*, P. Mellars and C. B. Stringer, eds., Edinburgh University Press, Edinburgh, pp.

Alimen, H. and A. Vignal, 1952, Etude statistique de bifaces acheuléens. Essai d'archéometrie, *Bulletin de la Société Préhistorique Française* **49**:56-72.

Allain, J. and J. Descouts, 1957, A propos d'une baguette à rainure armée de silex découverte dans le Magdalénien de Saint-Marcel., *L'Anthropologie* **61**:5-6:503-512.

Allain, J., 1979, L'Industrie lithique et osseuse de Lascaux, in: *Lascaux Inconnu*, A. Leroi-Gourhan and J. Allain, eds., C.N.R.S, Paris, pp. 87-102.

Allain, J., R. Desbrosse, J. Kozlowski, A. Rigaud, M. Jeannet et al., 1985, Le Magdalénien à navettes, *Gallia Préhistoire* **28**:37-124.

Altman, J., 1980, *Baboon Mothers and Infants*, Harvard University Press, Cambridge, MA.

Alvard, M. S., 2003, The adaptive nature of culture, *Evolutionary Anthropology* **12**:136-149.

Andrew, R. J., 1976, Use of formants in the grunts of baboons and other nonhuman primates, in: *Origins and Evolution of Language and Speech*, H. B. Steklis, S. R. Hamad and J. Lancaster, eds., New York, pp. 673-693.

Aoki, K., 1982, A condition for group selection to prevail over counteracting individual selection, *Evolution* **36**:832-842.

Arensburg, B., 1989, The hyoid bone from the Kebara 2 hominid, in: *Investigations in South Levantine Prehistory*, O. Bar-Yosef and B. Vandermeersch, eds., BAR International Series 497, Oxford, pp. 337-342.

Arensburg, B., A.-M. Tillier, B. Vandermeersch, H. Duday, L. A. Schepartz et al., 1989, A Middle Palaeolithic human hyoid bone, *Nature* **338**:758-760.

Arensburg, B., L. A. Schepartz, A.-M. Tillier, B. Vandermeersch and Y. Rak, 1990, A reappraisal of the anatomical basis for speech in Middle Palaeolithic hominids, *American Journal of Physical Anthropology* **83**:137-146.

Arthur, G., 1975, *An Introduction to the Ecology of Early Historic Communal Bison Hunting among the Northern Plains Indians*, Archaeological Survey of Canada. Mercury Series no. 37. National Musem of Man, Ottowa.

Arthur, W. B., 1999, Complexity and the economy, *Science* **284**:107-109.

Atran, S., 1982, Constraints on a theory of hominid tool-making behavior, *L'Homme* **22**:35-68.

Atran, S., 1990, *Cognitive Foundations of Natural History: Towards an Anthropology of Science*, Cambridge University Press, Cambridge.

Atran, S., 2003, Théorie cognitive de la culture, *L'Homme* **166**:107-144.

Atran, S. and A. Norenzayan, 2004, Religion's evolutionary landscape: Counterintuition, commitment, compassion, communion, *Behavioral and Brain Sciences* **27**:713-713-730.

Audouin, F. and H. Plisson, 1982, Les ocres et leur témoins au paléolithique en France : Enquête et expériences sur leur validité archéologique, *Cahiers du Centre de Recherches Préhistoriques de l'Université de Paris* **8**:33-80.

Avital, E. and E. Jablonka, 2000, *Animal Traditions: Behavioural Inheritance in Evolution*, Cambridge University Press, Cambridge.

Axelrod, R., 1984, *The Evolution of Cooperation*, Basic Books, New York.

Axelrod, R., 1986, An evolutionary approach to norms, *The American Political Science Review* **80**:1095-1111.

Axelrod, R. and D. Dion, 1988, The further evolution of cooperation., *Science* **242**:1385-1390.

Axelrod, R., 1997, *The Complexity of Cooperation: Agent-Based Models of Competition and Collaboration*, Princeton University Press, Princeton, NJ.

Babloyantz, A., 1986, *Molecules, Dynamics, and Life: An Introduction to the Self-Organization of Matter*, Wiley Interscience, New York.

Bächler, E., 1921, Das Drachenloch ob Vättis im Taminatal, 2445 m ü.M. und seine Bedeutung als paläontologische Fundstätte und prähistorische Niederlassung aus der Altsteinzeit (Paläolithikum) im Schweitzerlände, *Jahrbuch der St. Gallischen naturwissenschaflichen Gesellschaft* **57**:1-144.

Bächler, E., 1923, Die Forschungsergebnisse im Drachenloch ob Vättis im Taminatale 2445 m ü.M.: Nachtrag und Zusammenfassung, *Jahrbuch der St. Gallischen naturwissenschaftlichen Gesellschaft* **59**:79-118.

Bächler, E., 1940, *Das Drachenloch ob Vättis im Taminatale 2445 m ü.M*, Zollikofer und Cie, St. Gallen.

Barham, L. S. and P. Smart, 1996, An early date for the Middle Stone Age of central Zambia, *Journal of Human Evolution* **30**:287-290.

Barham, L. S., 2002, Systematic pigment use in the Middle Pleistocene of South-Central Africa, *Current Anthropology* **43**:181-190.

Bar-Yosef, O., 1988, Evidence for middle palaeolithic symbolic behaviour: A cautionary note., in: *L'Homme de Néandertal, Vol. 5, La Pensée*, O. Bar-Yosef, ed. E.R.A.U.L. 32., pp.

Bar-Yosef, O., H. Laville, L. Meignen, A.-M. Tillier, B. Vandermeersch et al., 1988, La Sépulture néanderthalienne de Kébara (Unité XII). in: *L'Homme de Neanderthal 5 (La Pensée)*, M. Otte, ed. pp. 17-24.

Bastiani, G., 1997, Results from the experimental manufacture of a bone flute with stone tools, in: *Mousterian "Bone Flute" and Other Finds from Divje Babe I Cave Site in Slovenia*, I. Turk, ed. Znanstvenoraziskovalni Center SAZU, Ljubljana, pp. 176-178.

Batson, C. D., 1991, *The Altruism Question: Toward a Social-Psychological Answer*, Lawrence Erlbaum Associates, Hillsdale, NJ.

Beaumont, P. B., 1973, Ancient pigment mines of southern Africa, *South African Journal of Science* **69**:140-146.

Beaumont, P. B., 1990a, Kathu, in: *Guide to Archaeological Sites in the Northern Cape*, P. B. Beaumont and D. Morris, eds., McGregor Museum, Kimberly, pp. 75-100.

Beaumont, P. B., 1990b, Nooitgedacht 2 and Roseberry Plain 1, in: *Guide to Archaeological Sites in the Northern Cape*, P. B. Beaumont and D. Morris, eds., McGregor Museum, Kimberly, pp. 4-6.

Bednarik, R. G., 1992, Palaeoart and Archaeological Myths, *Cambridge Archaeological Journal* **2**:27-57.

Bednarik, R. G., 1994, A taphonomy of palaeoart, *Antiquity* **68**.

Bednarik, R. G., 1995, Concept-mediated marking in the Lower Palaeolithic, *Current Anthropology* **36**:605-634.

Bednarik, R. G., 1998, The 'Australopithecine' cobble from Makapansgat, South Africa, *South African Archaeological Bulletin* **53**:4-8.

Bednarik, R. G., 2001, The oldest known rock art in the world, *Anthropologie (Brno)* **39**:89-97.

Bednarik, R. G., 2002, The oldest surviving rock art: A taphonomic review, *Origini* **24**:335-349.

Bednarik, R. G., 2003, A figurine from the African Acheulian, *Current Anthropology* **44**:405-412.

Beerli, P. and S. V. Edwards, 2002, When did Neanderthals and modern humans diverge? *Evolutionary Anthropology* **11**:60-63.

Behm-Blancke, G., 1987, Zur geistigen Welt des *Homo erectus* von Bilzingsleben., *Jahresschrift für mitteldeutsche Vorgeschichte* **70**:41-82.

Belfer-Cohen, A., 1989, The appearance of symbolic expression in the upper pleistocene of the Levant as compared to western Europe, in: *L'Homme de Néandertal, Vol. 5, La Pensée*, O. Bar-Yosef, ed. E.R.A.U.L. 32, pp. 25-30.

Belfer-Cohen, A. and E. Hovers, 1992, In the eye of the beholder: Mousterian and Natufian burials in the Levant., *Current Anthropology* **33**:463-473.

Benedict, R., 1934, *Patterns of Culture*, Houghton Mifflin, New York.

Berman, C. M., 1980, Early agonistic experience and rank acquisition among free-ranging infant rhesus monkeys, *International Journal of Primatology* **1**:153-170.

Binant, P., 1991, *Les Sépultures du Paléolithique.*, Archéologie aujourd'hui, Editions Errance, Paris.

Binford, L. R., 1981, *Bones: Ancient Men and Modern Myths.*, Academic Press, New York.

Blanc, A., 1957, A new paleolithic cultural element, probably of ideological significance: The clay pellets of the cave of the Basua (Savona), *Quaternaria* **4**:111-119.

Blehr, O., 1990, Communal hunting as a prerequisite for caribou (wild reindeer) as a human resource., in: *Hunters of the Recent Past*, L. B. Davis and B. O. K. Reeves, eds., Unwin Hyman, London, pp.

Boas, F., 1940, *Race, Language, and Culture*, MacMillan, New York.

Bocquet-Appel, J.-P. and J. L. Arsuaga, 1999, Age distributions of hominid samples at Atapuerca (SH) and Krapina could indicate accumulation by catastrophe, *Journal of Archaeological Science* **26**:327-338.

Bocquet-Appel, J.-P. and J.-L. Arsuaga, 2001, Probable catastrophic mortality of the Atapuerca (SH) and Krapina hominid samples, in: *Humanity from African Naissance to Coming Millenia*, P. V. Tobias, M. A. Raath, J. Moggi-Cecchi and G. A. Doyle, eds., Firenze University Press, Florence, pp. 149-158.

Boëda, E., J.-M. Geneste and C. Griggo, 1999, A Levallois point embedded in the vertebra of a wild ass (*Equus africanus*): Hafting, projectiles and Mousterian hunting weapons, *Antiquity* **73**:394-402.

Boehm, C., 1999a, The natural selection of altruistic traits, *Human Nature* **10**:205-253.

Boehm, C., 1999b, *Hierarchy in the Forest*, Harvard University Press, Cambridge, MA.

Boesch, C., 1991, Teaching among wild chimpanzees, *Animal Behavior* **41**:530-532.

Boesch, C., 1993, Aspects of transmission of tool-use in wild chimpanzees., in: *Tools, Language and Cognition in Human Evolution*, K. R. Gibson and T. Ingold, eds., Cambridge University Press, Cambridge, pp. 171-184.

Boesch, C., P. Marchesi, N. Marchesi, B. Fruth and F. Joulian, 1994, Is nut cracking in wild chimpanzees a cultural behaviour? *Journal of Human Evolution* **26**:325-338.

Boesch, C. and M. Tomasello, 1998, Chimpanzee and human cultures, *Current Anthropology* **39**:591-614.

Bonifay, E., 1962, Un ensemble rituel moustérien á la grotte de Régourdou (Montignac, Dordogne), *Proceedings of the 6th Congress of Prehistoric and Protohistoric Sciences, Rome* **2**:136-140.

Bonifay, E. and B. Vandermeersch, 1962, Dépôts rituels d'ossements d'ours dans le gisement moustérien du Régourdou (Montignac, Dordogne), *Comptes rendus hebdomadaires de l'Académie des Sciences* **255**:1635-1636.

Bonifay, E., 1989, Fréquence et signification des sépultures néandertaliennes., in: *L'Homme de Néandertal, Vol. 5, La Pensée*, O. Bar-Yosef, ed. E.R.A.U.L. 32, pp. 31-36.

Bonner, J. T., 1980, *The Evolution of Culture in Animals*, Princeton Science Library, Princeton University Press, Princeton.

Boorman, S. A. and P. R. Levitt, 1980, *The Genetics of Altruism*, Academic Press, New York.

Bordes, F., 1969, Os percé moustérien et os gravé acheuléen du Pêch de l'Azé II., *Quaternaria* **11**:1-6.

Borofsky, R., F. Barth, R. A. Shweder, L. Rodseth and N. M. Stolzenberg, 2001, WHEN: A converstion about culture, *American Anthropologist* **103**:432-446.

Boyd, R. R. and P. J. Richerson, 1982, Cultural transmission and the evolution of cooperative behavior, *Human Ecology* **10**:325-351.

Boyd, R. R. and P. J. Richerson, 1985, *Culture and the Evolutionary Process.*, University of Chicago Press, Chicago.

Boyd, R. R. and P. J. Richerson, 1992, Punishment allows the evolution of cooperation (or anything else) in sizeable groups, *Ethology and Sociobiology* **13**:171-195.

Boyd, R. R., H. Gintis, S. Bowles and P. J. Richerson, 2003, The evolution of altruistic punishment, *Proceedings of the National Academy of Sciences* **100**:3531-3535.

Brain, C. K., 1967, Hottentot food remains and their meaning in the interpretation of fossil bone assemblages, *Scientific Papers of the Namib Desert Research Station* **32**:1-11.

Brain, C. K., 1969, *The Contribution of the Namib Desert Hottentots to an Understanding of Australopithecine Bone Accumulations*, Scientific Papers of the Namib Desert Research Station 39.

Brain, C. K., 1976, Some principles in the interpretation of bone accumulations associated with man, in: *Human Origins*, G. L. Isaac and E. McCown, eds., Benjamin, Menlo Park, pp. 97-116.

Brain, C. K., 1981, *The Hunters or the Hunted?* The University of Chicago Press, Chicago.

Brandon, R. N. and R. M. Burian, eds., 1984, *Genes, Organisms, and Populations: Controversies over the Units of Selection*, MIT Press, Boston.

Breuil, H. and H. Obermaier, 1909, Crânes paléolithiques façonnés en coupes., *L'Anthropologie* **20**:523-530.

Breuil, H. and R. Lantier, 1959, *The Men of the Old Stone Age.*, St. Martin's Press, New York.

Brown, F. H. and I. McDougall, 1993, Geological age and setting, in: *The Nariokotome Homo erectus Skeleton*, A. Walker and R. E. F. Leakey, eds., Harvard University Press, Cambridge, pp. 9-20.

Bruner, E., G. Manzi and J. L. Arsuaga, 2003, Encephalization and allometric trajectories in the genus *Homo*: Evidence from the Neandertal and modern lineages, *Proceedings of the National Academy of Sciences* **100**:15335-15340.

Burns, T. R. and T. Dietz, 1992, Cultural evolution: Social rule systems, selection, and human agency, *International Sociology* **7**:259-283.

Burr, D. B., 1976a, Neandertal vocal tract reconstructions: a critical appraisal, *Journal of Human Evolution* **5**:285-290.

Burr, D. B., 1976b, Further evidence concerning speech in Neandertal man, *Man* **11**:104-110.

Butzer, K. W., 1980, Comment on "Red ochre and human evolution" by Ernst E. Wreschner, *Current Anthropology* **21**:635.

Byers, A. M., 1994, Symboling and the Middle-Upper Palaeolithic transition: A theoretical and methodological critique, *Current Anthropology* **35**:369-400.

Byers, A. M., 1999, Communication and material culture: Pleistocene tools as action cues, *Cambridge Archaeological Journal* **9**:23-41.

Byers, A. M., 2001, A pragmatic view of the emergence of Paleolithic symbol-using, in: *In the Mind's Eye : Multidisciplinary Approaches to the Evolution of Human Cognition*, A. Nowell, ed. International Monographs in Prehistory, Ann Arbor, pp. 50-62.

Byrne, R. W. and A. Whiten, eds., 1988, *Machiavellian Intelligence: Social Expertise and the Evolution of Intellect in Monkeys, Apes, and Humans.*, Clarendon Press, Oxford.

Campbell, D. T., 1965, Variation and selective retention in sociocultural evolution, in: *Social Change in Underdeveloped Areas: A Reinterpretation of Evolutionary Theory*, R. W. Mack, G. Blanksten and H. R. Barringer, eds., Schenkman, Cambridge, pp. 19-48.

Cantalupo, C. and W. D. Hopkins, 2001, Asymmetric Broca's area in great apes, *Nature* **414**:505.

Caramelli, D., C. Lalueza-Fox, V. Cristiano and et al., 2003, Evidence for a genetic discontinuity between Neandertals and 24,000-year-old anatomically modern Europeans, *Proceedings of the National Academy of Sciences* **100**:6593-6597.

Carbonell, E., M. Mosquera, A. Ollé and et al., 2003, Les premiers comportements funéraires auraient-ils pris place à Atapuerca, il y a 350,000 ans? *L'Anthropologie* **107**:1-14.

Carlisle, R. C. and M. Siegel, 1974, Some problems in the interpretation of Neanderthal speech capabilities: A reply to Lieberman, *American Anthropologist* **76**:319-322.

Carruthers, P., 2002, The cognitive functions of language, *Behavioral and Brain Sciences* **25**:(DRAFT).

Cavalli-Sforza, L. L. and M. W. Feldman, 1981, *Cultural Transmission and Evolution*, Princeton University Press, Princeton.

Chapais, B., 1995, Alliances as a means of competition in primates: evolutionary, *Yearbook of Physical Anthropology* **38, 1995**:115-136.

Chase, P. G. and H. L. Dibble, 1987, Middle Paleolithic symbolism: A review of current evidence and interpretations., *Journal of Anthropological Archaeology* **6**:263-296.

Chase, P. G., 1990, Sifflets du Paléolthique moyen (?). Les implications d'un coprolithe de coyote actuel, *Bulletin de la Société Préhistorique Française* **87**:165-167.

Chase, P. G., 1991, Symbols and Paleolithic artifacts: Style, standardization, and the imposition of arbitrary form, *Journal of Anthropological Archaeology* **10**:193-214.

Chase, P. G. and H. L. Dibble, 1992, Scientific archaeology and the origins of symbolism: A reply to Bednarik, *Cambridge Archaeological Journal* **2**:43-51.

Chase, P. G., 1994, Palaeolithic archaeology and the human mind: Comment on review feature *In Search of the Neanderthals: Solving the Puzzle of Human Origins*, by Christopher Stringer and Clive Gamble, *Cambridge Archaeological Journal* **4**:110-112.

Chase, P. G., 1995, Evidence for the use of bones as cutting boards in the French Mousterian, *Préhistoire Européenne* **7**:111-117.

Chase, P. G. and A. Nowell, 1998, Taphonomy of a suggested Middle Paleolithic bone flute from Slovenia, *Current Anthropology* **39**:549-553.

Chase, P. G., 1999, Symbolism as reference and symbolism as culture, in: *The Evolution of Culture: An Interdisciplinary View*, R. I. M. Dunbar, C. Knight and C. Power, eds., Edinburgh University Press, Edinburgh, pp. 34-49.

Chase, P. G., 2001a, 'Symbolism' is two different phenomena: Implications for archaeology and paleontology, in: *Humanity from African Naissance to Coming Millennia*, P. V. Tobias, M. A. Raath, J. Moggi-Cecchi and G. A. Doyle, eds., Firenze University Press, Florence, pp. 199-212.

Chase, P. G., 2001b, Multilevel information processing, archaeology, and evolution, in: *In the Mind's Eye : Multidisciplinary Approaches to the Evolution of Human Cognition*, A. Nowell, ed. Ann Arbor, pp. 121-136.

Chase, P. G., 2001c, Punctured reindeer phalanges from the Mousterian of Combe Grenal (France), *Arheološki Vestnik* **52**:17-23.

Chase, P. G., 2003, Comment on "The origin of modern human behavior" by Christopher S. Henshilwood and Curtis W. Marean, *Current Anthropology* **44**:637.

Cheney, D. L. and R. M. Seyfarth, 1990, *How Monkeys see the World*, University of Chicago Press, Chicago.

Chiarelli, B., 1987, The natural history of the ethical concept and its implications for the present natural equilibrium., *Human Evolution* **2**:413-418.

Chomsky, N., 1972, *Language and Mind*, Harcourt and Brace, New York.

Churchland, P. S. and T. Sejnowski, 1992, *The Computational Brain*, Computational neuroscience, MIT Press, Cambridge, MA.

Clark, G. A., 1989, Romancing the stones: Biases, style and lithics at La Riera., in: *Alternative Approaches to Lithic Analysis*, D. O. Henry and G. H. Odell, eds., Archeological Papers of the American Anthropological Association No. 1, Washington, D.C., pp. 27-50.

Clark, J. D., K. P. Oakley, L. H. Wells and J. A. C. McClelland, 1947, New Studies on Rhodesian Man, *Journal of the Royal Anthropological Institute* **77**:7-32.

Clark, J. D., K. D. Williamson, J. M. Michels and C. A. Marean, 1984, A Middle Stone Age occupation site at Porc Epic Cave, Dire Dawa (east-central Ethiopia). *African Archaeological Review* **2**:37-72.

Clark, J. D., Y. Beyene, G. WoldeGabriel, W. K. Hart, P. R. Renne et al., 2003, Stratigraphic, chronological and behavioural contexts of Pleistocene *Homo sapiens* from Middle Awash, Ethiopia, *Nature* **423**:747-752.

Close, A. E., 1980, The identification of style in lithic artefacts, *World Archaeology* **10**:223-236.

Close, A. E., 1989, Identifying style in stone artefacts: A case study from the Nile Valley., in: *Alternative Approaches to Lithic Analysis*, D. O. Henry and G. H. Odell, eds., Archaeological Papers of the American Anthropological Association, No. 1, pp. 3-26.

Collins, P., 2001, The celibacy of Shakers, in: *Celibacy, Culture and Society*, E. J. Sobo and S. Bell, eds., University of Wisconsin Press, Madison, WI, pp. 104-121.

Conkey, M. W., 1978, Style and information in cultural evolution: Toward a predictive model for the Paleolithic, in: *Social Archaeology: Beyond Subsistence and Dating*, C. L. Redman and et al., eds., Academic Press, New York, pp. 61-85.

Conkey, M. W., 1980, Context, structure, and efficacy in Paleolithic art and design, in: *Symbol as Sense: New Approaches to the Analysis of Meaning*, M. L. Foster and S. H. Brandes, eds., New York, pp. 225-248.

Conkey, M. W., 1990, Experimenting with style in archaeology, in: *The Uses of Style in Archaeology*, M. W. Conkey and C. Hastorf, eds., Cambridge University Press, Cambridge, pp. 1-17.

Cooke, C. K., 1963, Report on the excavations at Pomongwe and Tshangula Caves, Matopo Hills, Southern Rhodesia, *South African Archaeological Bulletin* 16:75-151.

Coolidge, F. L. and T. G. Wynn, 2004, A cognitive and neuropsychological perspective on the Châtelperronian, *Journal of Anthropological Research* 60:55-73.

Coolidge, F. L. and J. G. Wynn, 2005, Working memory, its executive functions, and the emergence of modern thinking, *Cambridge Archaeological Journal* 15:5-26.

Copeland, L., 1975, The Middle and Upper Paleolithic of Lebanon and Syria in light of recent research, in: *Problems in Prehistory: North Africa and the Levant*, F. Wendorf and A. E. Marks, eds., Southern Methodist University Press, Dallas, pp.

Cordwell, J. M., 1985, Ancient beginnings and modern diversity of the use of cosmetics, in: *The Psychology of Cosmetic Usage*, J. A. Graham and A. M. Kligman, eds., Praeger Scientific, New York, pp.

Corruccini, R. S., 1994, Reaganomics and the fate of the progressive Neanderthals, in: *Integrative Paths to the Past: Paleoanthropological Advances in Honor of F. Clark Howell*, R. S. Corruccini and R. L. Ciochon, eds., Prentice Hall, Englewood Cliffs, NJ, pp. 697-708.

Cox, S. J., T. J. Sluckin and J. Steele, 1999, Group size, memory and interaction rate in the evolution of cooperation, *Current Anthropology* 40:369-376.

Crelin, E. S., 1987, *The Human Vocal Tract : Anatomy, Function, Development, and Evolution*, Vintage Press, New York.

Crémades, M., 1996, L'Expression graphique au Paléolithique inférieur et moyen: l'exemple de l'Abri Suard (La Chaise-de-Vouthon, Charente), *Bulletin de la Société Préhistorique Française* 93:494-501.

Cremaschi, M., S. D. Lernia and E. A. A. Garcee, 1998, Some insights into the Aterian in the Libyan Sahara: Chronology, environment, and archaeology, *African Archaeological Review* 15:261-286.

Crompton, R. H. and J. A. J. Gowlett, 1993, Allometry and multidimensional form in Acheulean bifaces from Kilombe, Kenya, *Journal of Human Evolution* 25:175-200.

Daly, M., 1982, Some caveats about cultural transmission models, *Human Ecology* 10:401-408.

Dart, R. A., 1949, The predatory implemental technique of *Australopithecus*, *American Journal of Physical Anthropology* 7:1-38.

Dart, R. A., 1957, *The Osteodontokeratic Culture of Australopithecus prometheus*, Transvaal Museum Memoirs, no. 10, Pretoria.

Dart, R. A., 1958, The minimal bone-breccia content of Makapansgat and the Australopithecine predatory habit, *American Anthropologist* 60:923-931.

Dart, R. A., 1960, The bone tool-manufacturing ability of *Australopithecus prometheus*, *American Anthropologist* 62:134-143.

Dart, R. A., 1974, The waterworn Australopithecine pebble of many faces from Makapansgat, *South African Journal of Science* 70:167-169.

Datta, S. B., 1983, Relative power and the acquisition of rank, in: *Primate Social Relationships: An Integrated Approach*, R. A. Hinde, ed. Oxford University Press, Oxford, pp.

David, F. and P. Fosse, 1999, Le bison comme moyen de subsistance au Paléolithique: Gisements de plein air et sites en grotte, in: *Le Bison: Gibier et Moyen de Subsistance des Hommes du Paléolithique aux Paléoindiens des Grandes Plaines. Actes du Colloque International, Toulouse, 6-10 Juin 1995*, J.-P. Brugal, J. G. Enloe and J. Jaubert, eds., Editions APDCA, Antibes, pp. 120-141.

Davidson, I. and W. Noble, 1989, The archaeology of perception: Traces of depiction and language, *Current Anthropology* **30**:125-155.

Davidson, I., 1991, The archaeology of language origins: a review, *Antiquity* **65**:39-48.

Davidson, I. and W. Noble, 1992, Why the first colonisation of the Australian region is the earliest evidence of modern human behaviour, *Archaeology in Oceania* **27**:135-142.

Davidson, I. and W. Noble, 1993, Tools and language in human evolution, in: *Tools, Language and Cognition in Human Evolution*, K. R. Gibson and T. Ingold, eds., Cambridge University Press, Cambridge, pp. 363-388.

Davis, S., 1974, Incised bones from the Mousterian of Kebara Cave (Mount Carmel) and the Aurignacian of Ha-yonim (Western Galilee) Israel, *Paléorient* **2**:181-182.

Dawkins, R., 1976, *The Selfish Gene*, Oxford University Press, Oxford.

Dawkins, R., 1993, Viruses of the mind, in: *Dennett and His Critics: Demystifying Mind*, B. Dahlbom, ed. Blackwell, Cambridge, MA, pp. 13-27.

de Lumley, H., 1966, Les fouilles de Terra Amata à Nice. Premiers résultats., *Bulletin du Musée d'Anthropologie et Préhistoire de Monaco* **13**:29-51.

de Waal, F. B. M., 1982, *Chimpanzee Politics: Power and Sex among Apes*, John Hopkins University Press, Baltimore.

de Waal, F. B. M., 1989, *Peacemaking among Primates*, Harvard University Press, Cambridge, MA.

de Waal, F. B. M., 1996, *Good Natured: The Origins of Right and Wrong in Humans and Other Animals*, Harvard University Press, Cambridge, MA.

Deacon, H. J., 1995, Two Late Pleistocene-Holocene archaeological depositories from the Southern Cape, South Africa, *South African Archaeological Bulletin* **50**:211-231.

Deacon, H. J., 2001, Modern human emergence: An African archaeological perspective, in: *Humanity from African Naissance to Coming Millenia*, P. V. Tobias, M. A. Raath, J. Moggi-Cecchi and G. A. Doyle, eds., Firenze University Press, Florence, pp. 213-223.

Deacon, T. W., 1992, The neural circuitry underlying primate calls and human language, in: *Language Origin: A Multidisciplinary Approach*, J. Wind, B. H. Bichakjian, A. Nocentini and B. Chiarelli, eds., Kluwer Academic Publishers, Dordrecht, pp. 121-162.

Deacon, T. W., 1997, *The Symbolic Species: The Co-evolution of Language and the Brain*, W.W. Norton and Company, New York.

Debénath, A., J.-P. Raynal, J. Roche, J.-P. Texier and D. Ferembach, 1986, Stratigraphie, habitat, typologie et devenir de l'Atérien marocain: données récentes, *L'Anthropologie* **90**:233-246.

Debénath, A., 1994, L'Atérien du Nord de l'Afrique et du Sahara, *Sahara* **6**:21-30.

Debénath, A. and H. L. Dibble, 1994, *The Handbook of Paleolithic Typology.Vol I. The Lower and Middle Paleolithic of Europe*, The University Museum Press. University of Pennsylvania, Philadelphia.

Debénath, A., P. G. Chase and H. L. Dibble, 2002, A propos d'un poinçon provenant de la grotte des Fées à Châtelperron, *Bulletin de la Société Préhistorique Française* **99**:378-379.

Defleur, A., O. Dutour, H. Valladas and B. Vandermeersch, 1993, Cannibals among the Neanderthals? *Nature* **362**:214.

Defleur, A., T. D. White, P. Valensi, L. Slimak and É. Crégut-Bonnoure, 1999, Neanderthal Cannibalism at Moula-Guercy, Ardèche, France, *Science* **286**:128-131.

DeGusta, D., W. H. Gilbert and S. P. Turner, 1999, Hypoglossal canal size and hominid speech, *Proceedings of the National Academy of Sciences* **96**:1800-1804.

Dennett, D. C., 1987, *The Intentional Stance*, Bradford Books MIT Press, Cambridge.

Dennett, D. C., 1991, *Consciousness Explained*, Allan Lane, London.

Dennett, D. C., 1995, *Darwin's Dangerous Idea: Evolution and the Meanings of Life*, Touchstone, Simon and Schuster, New York.

D'Errico, F., C. Gaillard and V. N. Misra, 1989, Collection of non-utilitarian objects by *Homo erectus* in India, in: *Hominidae. Proceedings of the Second International Congress of Human Palaeontology, Turin*, G. Giacobini, ed. Editoriale Jaca Book, Milan, pp. 237-239.

D'Errico, F. and P. Villa, 1997, Holes and grooves: the contribution of microscopy and taphonomy to the problem of art origins, *Journal of Human Evolution* **33**:1-31.

D'Errico, F., P. Villa, A. C. Pinto Llona and R. Ruiz Idarraga, 1998, A Middle Palaeolithic origin of music? Using cave-bear accumulations to assess the Divje Babe I bone 'flute', *Antiquity* **72**:65-79.

D'Errico, F., J. Zilhão, M. Julien, D. Baffier and J. Pelegrin, 1998, Neanderthal acculturation in Western Europe? A critical review of the evidence and its interpretation, *Current Anthropology* **39**:s1-s44.

D'Errico, F. and A. Nowell, 2000, A new look at the Berekhat Ram figurine: Implications for the origins of symbolism, *Cambridge Archaeological Journal* **10**:123-167.

D'Errico, F., C. S. Henshilwood and P. Nilssen, 2001, An engraved bone fragment from 70,000-year-old Middle Stone Age levels at Blombos Cave, South Africa: implications for the origin of symbolism and language, *Antiquity* **75**:309-3098.

D'Errico, F. and M. Soressi, 2002, Systematic use of manganese pigment by Pêch-de-l'Azé Neandertals: Implications for the origin of behavioral modernity (Paper presented to the Paleoanthropological Society Meetings, Denver, 2002), *Journal of Human Evolution* **42**:A13.

Deutsch, M., 1949, A theory of co-operation and competition, *Human Relations* **2**:129-152.

Dibble, H. L., 1987, The interpretation of Middle Paleolithic scraper morphology, *American Antiquity* **52**:109-117.

Dibble, H. L., 1989, The Implications of stone tool types for the presence of language during the Middle Paleolithic, in: *In The Human Revolution: Behavioural and Biological Perspectives on the Origins of Modern Humans*, P. Mellars and C. B. Stringer, eds., Edinburgh University Press, Edinburgh, pp. 415-432.

Dibble, H. L. and P. G. Chase, 1993, On Mousterian and Natufian burials in the Levant, *Current Anthropology* **34**:170-172.

Dibble, H. L., 1995, Middle Paleolithic scraper reduction: background, clarification, and review of evidence to data, *Journal of Archaeological Method and Theory* **2**:299-368.

Donald, M., 1991, *Origins of the Modern Mind: Three Stages in the Evolution of Culture and Cognition*, Harvard University Press, Cambridge.

Du Brul, E. L., 1976, Biomechanics of speech sounds, in: *Origins and Evolution of Language and Speech*, H. B. Steklis, S. R. Harnad and J. Lancaster, eds., New York, pp. 631-642.

Duchin, L. E., 1990, The evolution of articulate speech: Comparative natomy of the oral cavity in *Pan* and *Homo*, *Journal of Human Evolution* **19**:687-697.

Duff, A. I., G. A. Clark and T. J. Chadderdon, 1992, Symbolism in the Early Paleolithic: A conceptual odyssey., *Cambridge Archaeological Journal* **2**:211/229.

Dugatkin, L., 1999, *Cheating Monkeys and Citizen Bees: The Nature of Cooperation in Animals and Humans*, The Free Press, New York.

Dunbar, R. I. M., 1988, *Primate Social Systems*, Cornell University Press, Ithaca, NY.

Dunbar, R. I. M., 1998, The social brain hypothesis, *Evolutionary Anthropology* **6**:178-186.

Dunnell, R. C., 1978, Style and function: a fundamental dichotomy., *American Antiquity* **4**:192-202.

Dunnell, R. C., 1980, Evolutionary theory and archaeology, *Advances in Archaeological Method and Theory* **3**:35-99.

Durham, W. H., 1990, Advances in evolutionary culture theory, *Annual Review of Anthropology* **19**:187-210.

Durham, W. H., 1991, *Coevolution: Genes, Culture, and Human Diversity*, Stanford University Press, Stanford.

Durkheim, E., 1915 [1965], *The Elementary Forms of the Religious Life*, J. W. Swain, Free Press, New York.

Durkheim, E., 1938 [1964], *The Rules of Sociological Method, Eighth Edition*, S. A. S. a. J. H. Mueller, The Free Press, New York.

Ehrenberg, K., 1953, Die paläontologische, prähistorische und paläo-ethnologische Bedeutung der Salzofenhöhle im Lichte der letzten Forschungen, *Quartär* **6**:19-51.

Eve, R. A., S. Horsfall and M. E. Lee, 1997, *Chaos, Complexity and Sociology: Myths Models and Theories*, Sage.

Facchini, F., 1998, Il simbolismo nell'uomo preistorico: Aspetti ermeneutici e manifestazioni, *Rivista di Scienze Preistoriche* **49**:651-671.

Falk, D., 1975, Comparative anatomy of the larynx in man and the chimpanzee: Implications for language in Neanderthal, *American Journal of Physical Anthropology* **43**:123-132.

Falk, D., 1980a, A re-analysis of the South African Australopithecine natural endocasts, *American Journal of Physical Anthropology* **53**:525-540.

Falk, D., 1980b, Language, handedness, and primate brains: Did the Australopithecines sign? *American Anthropologist* **82**:72-78.

Falk, D., 1983, Cerebral cortices of East African early hominids, *Science* **221**:1072-1074.

Falk, D., 1985, Hadar AL 162-28 endocast as evidence that brain enlargement preceded cortical reorganization in hominid evolution, *Nature* **313**:45-47.

Falk, D., 1986, Endocast morphology of Hadar Hominid AL 162-28 (Reply to Holloway and Kimbel), *Nature* **321**:536-537.

Falk, D., J. C. Redmond and et al., 1999, Early hominid brain evolution: A new look at old endocasts, *Journal of Human Evolution* **38**:695-717.

Farizy, C., 1988, Sépultures et croyances. La culture matérielle, in: *De Néandertal à Cro-Magnon.*, C. Farizy, ed. Musée de Préhistoire de l'Ile-de-France, Nemours, pp. 54-55.

Farizy, C., F. David and J. Jaubert, 1994, *Hommes et Bisons du Paléolithique moyen à Mauran (Haute-Garonne)*, Gallia Préhistoire, Supplément 30, Centre National de la Recherche Scientifique, Paris.

Feraud, G., D. York, C. M. Hall, N. Goren-Inbar and H. P. Schwarcz, 1983, 40Ar/39Ar age limit for an Acheulean site in Israel, *Nature* **304**:263-265.

Fernández-Jalvo, Y., J. C. Díez, I. Cáceres and J. Rosell, 1999, Human cannibalism in the Early Pleistocene of Europe (Gran Dolina, Sierra de Atapuerca, Burgos, Spain), *Journal of Human Evolution* **37**:591-622.

Field, A. J., 2001, *Altruistically Inclined? The Behavioral Sciences, Evolutionary Theory, and the Origins of Reciprocity*, University of Michigan Press, Ann Arbor.

Fisher, J. and R. A. Hinde, 1949, The opening of milk bottles by birds, *British Birds* **42**:347-357.

Fitch, W. T., 2000, The evolution of speech: A comparative review, *Trends in Cognitive Sciences* **4**:258-267.

Flinn, M. V. and R. D. Alexander, 1982, Culture theory: The developing synthesis from biology, *Human Ecology* **10**:383-400.

Fodor, J. A., 1975, *The Language of Thought*, Harvard University Press, Cambridge, MA.

Fodor, J. A., 1983, *The Modularity of the Mind*, MIT Press, Cambridge.

Foley, R. A. and P. C. Lee, 1991, Ecology and energetics of encephalization in hominid evolution, *Philosophical Transactions of the Royal Society, London B* **334**:223-232.

Formicola, V., 1988, Interproximal grooving of teeth: Additional evidence and interpretation, *Current Anthropology* **29**:663-671.

Fouts, R. S. and R. L. Budd, 1979, Artificial and human language acquisition in a chimpanzee, in: *The Great Apes*, D. A. Hamburg and E. R. McCown, eds., Benjamin Cummings, Menlo Park, CA, pp. 375-392.

Fouts, R. S., A. Hirsch and D. Fouts, 1982, Cultural transmission of a human language in a chimpanzee mother and infant, in: *Child Nurturance*, H. Fitzgerald, J. Mullins and P. Gage, eds., Plenum Press, Gage, NY, Vol._3, pp.

Fouts, R. S., 1987, Chimpanzee signing and emergent levels, in: *Cognition, Language, and Consciousness: Integrative Levels*, G. Greenberg and E. Tobach, eds., Erlbaum Associates, Hillsdale, NJ, pp. 57-84.

Fox, R. G. and B. J. King, 2002, Introduction: Beyond culture worry, in: *Anthropology beyond Culture*, R. G. Fox and B. J. King, eds., Berg, Oxford, pp. 1-22.

Frank, R. H., 1988, *Passions within Reason*, Norton, New York.

Fridrich, J., 1975, Ein Beitrag zur Frage nach den Anfängen des künstlerischen Sinnes und Urmenschen (Vor-Neanderthal), *Pamatky Archeologicke* **68**:5.

Frison, G. C., 1968, A functional analysis of certain chipped stone tools, *American Antiquity* **33**:149-155.

Gallistel, C. R., 1989, Animal cognition: representation of space, time, and number, *Annual Review of Psychology* **40**:155-189.

Gardner, B. T. and R. A. Gardner, 1978, Comparative psychology and language acquisition, in: *Psychology: the State of the Art*, K. Salzinger and F. L. Denmark, eds., pp. 37-76.

Gardner, R. A. and B. T. Gardner, 1969, Teaching sign language to a chimpanzee, *Science* **165**:213-224.

Gargett, R. H., 1989, Grave shortcomings: The evidence for Neandertal burial, *Current Anthropology* **30**:157-190.

Gargett, R. H., 1999, Middle Paleolithic burial is not a dead issue: The view from Qafzeh, Saint-Césaire, Kebara, Amud, and Dederiyeh, *Journal of Human Evolution* **37**:27-90.

Gargett, R. H., 2000, A response to Hovers, Kimbel and Rak's argument for the purposeful burial of Amud 7 *Journal of Human Evolution* **39**:261-266

Gautier, A., 1986, Une histoire de dents: les soi-disant incisives travaillées du Paléolithique moyen de Sclayn, *Helinium* **26**:177-181.

Geertz, C., 1973, *The Interpretation of Cultures*, Basic Books, New York.

Gellner, E., 1988, Origins of society, in: *Origins: The Darwin College Lectures*, A. C. Fabian, ed. Cambridge University Press, Cambridge, pp. 128-140.

Geschwind, N., 1970, Organization of language and the brain, *Science* **170**:940-944.

Gibson, K. R., 1983, Comparative neurobehavioral ontogeny and the constructionist approach to the evolution of the brain, object manipulation and language, in: *Glossogenetics: The Origin and Evolution of Language*, E. de Grolier, ed. Harwood Academic Publishers, London, pp. 37-62.

Gibson, K. R., 1988, Brains size and the evolution of language, in: *The Genesis of Language: A Different Judgment of the Evidence.*, M. E. Landsberg, ed. Mouton de Gruyter, Berlin, pp. 149-172.

Gibson, K. R., 1990, New perspectives on instincts and intelligence: Brain size and the emergence of hierarchical mental constructional skills, in: *Language" and Intelligence in Monkeys and Apes*, S. T. Parker and K. R. Gibson, eds., Cambridge University Press, Cambridge, pp. 97-128.

Gibson, K. R., 1993, Tool use, language and social behavior in relationship to information processing capacities, in: *Tools, Language and Cognition in Human Evolution.*, K. R. Gibson and T. Ingold, eds., Cambridge University Press, Cambridge, pp. 251-270.

Gibson, K. R., 1996, The ontogeny and evolution of the brain, cognition, and language, in: *Handbook of Human Symbolic Evolution*, A. Lock and C. R. Peters, eds., Clarendon Press, Oxford, pp. 407-431.

Gibson, K. R., 2002, Customs and cultures in animals and humans: Neurobiological and evolutionary considerations, *Anthropological Theory* 2:323-339.

Giesen, B., 1991, Code, process and situation in cultural selection, *Cultural Dynamics* 4:173-185.

Goodall, J., 1986, *The Chimpanzees of Gombe: Patterns of Behavior*, The Belknap Press of Harvard University Press, Cambridge.

Goodenough, O., 1995, Mind viruses: Culture, evolution and the puzzle of altruism, *Social Science Information* 34:287-320.

Goodenough, W. H., 1981, *Culture, Language, and Society (Second Edition)*, T. Parsons, Benjamin/Cummings Publishing Co, Menlo Park, CA.

Goodwin, B., 1994, *How the Leopard Changed Its Spots: The Evolution of Complexity*, Charles Scribner's Sons, New York.

Gordon, B., 1990, World *Rangifer* communal hunting, in: *Hunters of the Recent Past*, L. B. Davis, ed. Unwin Hyman, London, pp.

Goren-Inbar, N., 1986, A figurine from the Acheulian site of Berekhat Ram, *Mitekufat Haeven* 19:7-12.

Goren-Inbar, N. and S. Peltz, 1995, Additional Remarks on the Berekhat Ram figurine, *Rock Art Research* 12:131-132.

Gould, S. J. and R. C. Lewontin, 1979, The spandrels of San Marco and the Panglossian paradigm: A critique of the adaptationist programme, *Proceedings of the Royal Society of London* **B 205**:581-598.

Gowlett, J. A. J., 1984a, *Ascent to Civilization*, Collins, London.

Gowlett, J. A. J., 1984b, Mental abilities of early man: A look at some hard evidence, in: *Hominid Evolution and Community Ecology: Prehistoric Human Adaptation in Biological Perspective*, R. A. Foley, ed. Academic Press, London, pp. 167-192.

Gowlett, J. A. J. and R. H. Crompton, 1994, Kariandusi: Acheulian morphology and the question of allometry, *The African Archaeological Review* 12:3-42.

Gowlett, J. A. J., 1996, Mental abilities of early Homo: Elements of constraint and choice in rule systems, in: *Modelling the Early Human Mind*, P. Mellars and K. R. Gibson, eds., Cambridge, pp. 191-216.

Greenfield, P. M. and E. S. Savage-Rumbaugh, 1991, Imitation, grammatical development, and the invention of protogrammar by an ape, in: *Biological and Behavioral Determinants of Language and Development*, N. A. Krasnegor, D. M. Rumbaugh, R. L. Schieffelbusch and M. Studdert-Kennedy, eds., Lawrence Erlbaum Associates, Hillsdale, NJ, pp. 235-258.

Gruet, M., 1959, Amoncellement pyramidal de sphères calcaires dans une source fossil moustérienne à El-Guettar (Sud Tunisien), *Actes du 2e Congrès Panafricain de Préhistoire, Alger* 449-456.

Guilmet, G. M., 1977, The evolution of tool-using and tool-making behaviour, *Man* **12**:33-47.

Gutiérrez, G., D. Sánchez and A. Marín, 2002, A reanalysis of the ancient mitochondrial DNA sequences recovered from Neanderthal bones, *Molecular Biology and Evolution* **19**:1359-1366.

Hamilton, W. D., 1964, The genetical evolution of social behavior I, II, *Journal of Theoretical Biology* **7**:1-52.

Hannerz, U., 1992, *Cultural Complexity : Studies in the Social Organization of Meaning*, Columbia University Press, New York.

Hanson, F. A., 2004, The new superorganic, *Current Anthropology* **45**:467-482.

Harms, W., 1996, Cultural evolution and the variable phenotype, *Biology and Philosophy* **11**:357-375.

Harrold, F. B., 1980, A comparative analysis of Eurasian Paleolithic burials, *World Archaeology* **12**:195-211.

Hayden, B., 1993, The cultural capacities of Neandertals: a review and reevaluation, *Journal of Human Evolution* **24**:113-146.

Hayes, K. J. and C. Hayes, 1951, The intellectual development of a home-raised chimpanzee, *Proceedings of the American Philosophical Society* **95**:105-109.

Henri-Martin, G., 1961, Le niveau de Châtelperron à la Quina (Charente), *Bulletin de la Société Préhistorique Française* **58**:796-808.

Henri-Martin, G., 1969, La Quina, in: *Berry - Poitou -Charentes. Livret-Guide de l'Excursion A4. UISPP 1976*, J.-P. Rigaud and B. Vandermeersch, eds., 158-62, pp.

Henri-Martin, G., 1976, La Quina, commune de Gardes, Charente, in: *Livret-Guide de L'excursion A 4, Sud-Ouest (Aquitaine et Charente), IXe Congrès de l'UISPP, Nice 13-18 septembre 1976*, J.-P. Rigaud and B. Vandermeersch, eds., Louis Jean, Gap, pp.

Henshilwood, C. S. and J. Sealy, 1997, Bone artefacts from the Middle Stone Age at Blombos Cave, Southern Cape, South Africa, *Current Anthropology* **38**:890-896.

Henshilwood, C. S., J. C. Sealy, R. J. Yates and et al., 2001, Blombos Cave, Southern Cape, South Africa: Preliminary report on the 1992-1999 excavations of the Middle Stone Age levels, *Journal of Archaeological Science* **28**:421-448.

Henshilwood, C. S., F. D'Errico, R. Yates, Z. Jacobs, C. Tribolo et al., 2002, Emergence of modern human behavior: Middle Stone Age engravings from South Africa, *Science* **295**:1278-1280.

Henshilwood, C. S., F. D'Errico, M. Vanhaeren, K. van Niekerk and Z. Jacobs, 2004, Middle Stone Age shell beads from South Africa, *Science* **304**:404.

Herrnstein, R. J., D. H. Loveland and C. Cable, 1985, Natural concepts in pigeons, in: *Directed Readings for the Principles of Learning and Behavior, 2nd edition*, N. A. Marler, ed. Brooks/Cole, Monterey, CA, pp. 214-230.

Hewes, G. W., 1973, An explicit formulation of the relationship between tool-using, tool-making, and the emergence of language, *Visible Language* **7**:100-127.

Hewes, G. W., 1975, *Language Origins: A Bibliography*, Mouton, The Hague.

Hill, W. W., 1944, The Navaho Indians and the Ghost Dance of 1890, *American Anthropologist* **46**:523-527.

Hinde, R. A., ed. 1983, *Primate Social Relationships: An Integrated Approach*, Oxford University Press, Oxford.

Hockett, C., 1959, Animal "languages" and human language, in: *The Evolution of Man's Capacity for Culture*, J. N. Spuhler, ed. Wayne State University Press, Detroit, pp. 32-39.

Hockett, C. and R. Ascher, 1964, The human revolution, *Current Anthropology* **5**:135-168.

Hockett, C. and S. Altmann, 1968, A note on design features, in: *Animal Communication*, T. A. Sebeok, ed. Indiana University Press, Bloomington, pp. 574-575.

Hoffecker, J. F., G. Baryshnikov and O. Potapova, 1991, Vertebrate remains from the Mousterian site of Il'skaya I (Northern Caucasus, U.S.S.R.): New analysis and interpretation, *Journal of Archaeological Science* **18**:113-147.

Hohman, G. and B. Fruth, 2003, Culture in bonobos? Between-species and within-species variation in behavior, *Current Anthropology* **44**:563-571.

Holloway, R. L., 1966, Cranial capacity, neural reorganization, and hominid evolution: A search for more suitable parameters, *American Anthropologist* **68**:103-121.

Holloway, R. L., 1969, Culture: A human domain, *Current Anthropology* **10**:395-412.

Holloway, R. L., 1972, Fossil evidence of the evolution of the human brain, *American Journal of Physical Anthropology* **37**:173-186.

Holloway, R. L., 1973, Endocranial volumes of early African hominids and the role of brain in human mosaic evolution, *Journal of Human Evolution* **2**:449-459.

Holloway, R. L., 1975, Early hominid endocasts: Volumes, morphology, and significance for hominid evolution, in: *Primate Functional Morphology and Evolution*, R. Tuttle, ed. Mouton, The Hague, pp. 393-415.

Holloway, R. L., 1976, Paleoneurological evidence for language origins, in: *Origins and Evolution of Language and Speech.*, S. R. Harnad and M. Stekelis, eds., pp. 330-348.

Holloway, R. L., 1979, Brain size, allometry and reorganization: Toward a synthesis, in: *Development and Evolution of Brain Size: Behavioral Implications*, M. E. Hahn, C. Jensen and B. C. Dudeck, eds., Academic Press, New York, pp. 59-89.

Holloway, R. L., 1981a, Revisiting the South African Taung Australopithecine endocast: The position of the lunate sulcus as determined by the stereoplotting technique, *American Journal of Physical Anthropology* **56**:43-58.

Holloway, R. L., 1981b, The Indonesian *Homo erectus* brain casts revisited, *American Journal of Physical Anthropology* **55**:503-521.

Holloway, R. L., 1981c, Culture, symbols, and human brain evolution, *Dialectical Anthropology* **5**:287-303.

Holloway, R. L. and M. C. de la Coste-Lareymondie, 1982, Brain endocast asymmetry in pongids and hominids: some preliminary findings on the palaeontology of cerebral dominance, *American Journal of Physical Anthropology* **58**:101-110.

Holloway, R. L., 1983a, Human brain evolution: A search for units, models and synthesis, *Canadian Journal of Anthropology* **3**:215-230.

Holloway, R. L., 1983b, Human paleontological evidence relevant to language behavior, *Human Neurobiology* **2**:105-114.

Holloway, R. L., 1983c, Cerebral brain endocast pattern of *Australopithecus afarensis* hominid, *Nature* **303**:420-422.

Holloway, R. L., 1985, The poor brain of *Homo sapiens neanderthalensis:* See what you please., in: *Ancestors: The Hard Evidence*, E. Delson, ed. Alan R. Liss, Inc, New York, pp. 319-324.

Holloway, R. L. and W. H. Kimbel, 1986, Endocast morphology of Hadar hominid AL 162-28, *Nature* **321**:536.

Holloway, R. L., D. C. Broadfield, M. S. Yuan and P. V. Tobias   2003   The lunate sulcus and early hominid brain evolution: Toward the end of a controversy. In *Abstracts of the 73rd Annual Meetings of the American Association of Physical Anthropologists*. Transl, Tempe, AZ.

Holloway, R. L., D. C. Broadfield and M. S. Yuan, 2004, *Brain Endocasts – The Paleoneurological Evidence*, vol. 3, Translated by The Human Fossil Record., Wiley-Liss, Hoboken, NJ.

Hörmann, K., 1933, *Die Petershöhle bei Velden in Mittelfranken: Eine altpaläolitische Station*, Nurenberg.

Horrocks, J. and W. Hunte, 1983, Maternal rank and offspring rank in vervet monkeys: An appraisal of the mechanisms of rank acquisition, *Animal Behavior* **31**:772-782.

Horusitzky, F. Z., 2003, Les flûtes paléolithiques: Divje Babe I, Istállóskö, Lokve, etc. Point de vue des experts et des contestataires, *Arheološki Vestnik* **54**:45-66.

Houghton, P., 1993, Neandertal supralaryngeal vocal tract, *American Journal of Physical Anthropology* **90**:139-146.

Hovers, E., B. Vandermeersch and O. Bar-Yosef, 1997, A Middle Palaeolithic engraved artefact from Qafzeh Cave, Israel, *Rock Art Research* **14**:79-87.

Hovers, E., W. H. Kimball and Y. Rak, 2000, The Amud 7 skeleton – still a burial. Response to Gargett, *Journal of Human Evolution* **39**:253-260.

Howell, F. C., 1960, European and northwest African Middle Pleistocene hominids, *Current Anthropology* **1**:195-232.

Howell, F. C., 1966, Observations on the earlier phases of the European Lower Paleolithic, *American Anthropologist* **68**:88-201.

Hublin, J.-J., F. Spoor, M. Braun, F. Zonneveld and S. Condemi, 1996, A late Neanderthal associated with Upper Palaeolithic artefacts, *Nature* **381**:224-226.

Hull, D., 1981, Units of evolution: A metaphysical essay, in: *The Philosophy of Evolution*, U. L. Jensen and R. Harré, eds., Harvester Press, Brighton, pp.

Hutchins, E., 1995, *Cognition in the Wild*, MIT Press, Cambridge, MA.

Ingold, T., 1994, Tool-using, tool-making, and the evolution of language, in: *Hominid Culture in Primate Perspective.*, D. Quiatt and J. Itani, eds., University Press of Colorado, Niwor, CO, pp. 279-314.

Isaac, G. L., 1976a, Stages of cultural elaboration in the Pleistocene: Possible archaeological indicators of the development of language capabilities, *Annals of the New York Academy of Sciences* **280**:275-288.

Isaac, G. L., 1976b, The activities of early African hominids: A review of archaeological evidence from the time span two and a half to one million years ago, in: *Human Origins: Louis Leakey and the East African Evidence.*, G. L. Isaac and E. McCown, eds., W.A. Benjamin, Menlo Park, pp. 483-514.

Isaac, G. L., 1978, The food-sharing behavior of protohuman hominids, *Scientific American* **238**:90-108.

Isaac, G. L., 1983, Bones in contention: Competing explanations for the juxtaposition of Early Pleistocene artifacts and faunal remains, in: *Animals and Archaeology*, J. Clutton-Brock and C. Grigson, eds., British Archaeological Reports International Series 163, Oxford, pp. 3-20.

Itani, J. and A. Nishimura, 1973, The study of infrahuman culture in Japan: A review, in: *Precultural Human Behavior: Symposia of the Fourth Congress of the International Primatological Society, Volume 1*, E. W. Menzel, ed. S. Karger, Basel, pp. 26-50.

Jantsch, E., 1980, *The Self-Organizing Universe: Scientific and Human Implications of the Emerging Paradigm of Evolution*, Pergamon Press, Oxford.

Jaubert, J., M. Lorblanchet, H. Laville, R. Slott-Moller, A. Turq et al., 1990, *Les Chasseurs d'Aurochs de La Borde. Un Site du Paléolithique Moyen (Livernon, Lot)*, Documents d'Archéologie Française, No. 27, Editions de la Maison des sciences de l'Homme, Paris.

Jelinek, A. J., 1976, Form, function and style in lithic analysis, in: *Cultural Change and Continuity: Essays in Honor of James Bennett Griffin*, C. E. Cleland, ed. Academic Press, New York, pp. 19-33.

Jelinek, A. J., 1981, The Middle Paleolithic in the Southern Levant from the perspective of the Tabun Cave, in: *Préhistoire du Levant: Chronologie et organisation de l'espace depuis les origines jusqu'au VIème millénnaire*, J. Cauvin and P. Sanlaville, eds., Centre National de la Recherche Scientifique, Paris, pp. 265-280.

Jéquier, J.-P., 1975, *Le Moustérien alpin*, Eburodunum II, Institut d'Archéologie Yverdonoise, Yverdon.

Jerison, H. J., 1976, Discussion paper: The paleoneuraology of language, in: *Origins and Evolution of Language and Speech*, H. B. Steklis, S. R. Harnad and J. Lancaster, eds., pp. 370-382.

Jerison, H. J., 1982, The evolution of biological intelligence, in: *Handbook of Human Intelligence*, R. J. Sternberg, ed. Cambridge University Press, Cambridge, pp. 723-791.

Johnson, L. L., 1978, A history of fint-knapping experimentation, *Current Anthropology* **19**:337-372.

Kauffman, S. A., 1995, *At Home in the Universe: The Search for Laws of Self-Organization and Complexity*, Oxford University Press, Oxford.

Kaufman, D., 2002, Redating the social revolution: The case for the Middle Paleolithic, *Journal of Anthropological Research* **58**.

Kaufman, S. A., 1993, *The Origins of Order: Self-Organization and Selection in Evolution*, Oxford University Press, Oxford.

Kawai, M., 1965, Newly-acquired pre-cultural behavior of the natural troop of Japanese monkeys on Koshima Islet, *Primates* **6**:1-30.

Kawamura, S., 1959, The process of sub-culture propagation among Japanese macaques, *Primates* **2**:43-60.

Kay, R. F., M. Cartmill and M. Balow, 1998, The hypoglossal canal and the origin of human vocal behavior, *Proceedings of the National Academy of Sciences* **95**:5417-5419.

Keeley, L. H., 1980, *Experimental Determination of Stone Tool Uses*, University of Chicago Press, Chicago.

Keeley, L. H., 1982, Hafting and retooling: Effects on the archaeological record, *American Antiquity* **47**:798-809.

Keith, A., 1948, *A New Theory of Human Evolution*, Watts, London.

Keller, L., ed. 1998, *Levels of Selection in Evolution*, Princeton University Press, Princeton.

Kellogg, W. N. and L. A. Kellogg, 1933, *The Ape and the Child*, McGraw-Hill, New York.

Kelsall, J. P., 1968, *The Migratory Barren-Ground Caribou of Canada*, Department of Indian Affairs and Northern Development, Canadian Wildlife Service, Ottawa.

King, B. J., 1999, New directions in primate social learning, in: *Mammalian Social Learning*, K. R. Gibson and H. O. Box, eds., Cambridge University Press, Cambridge, pp.

Kitahara-Frisch, J., 1978, Stone tools as indicators of linguistic ability in early man, *Kagaku Kisoron Gakkai Annals* **5**:101-109.

Klein, R. G., 1973, Geological antiquity of Rhodesian man, *Nature* **244**:311-312.

Klein, R. G., 1979, Stone age exploitation of animals in southern Africa, *American Scientist* **67**:151-160.

Klein, R. G., K. Allwarden and C. Wolf, 1983, The calculation and interpretation of ungulate age profiles from dental crown heights, in: *Hunter-Gatherer Economy in Prehistory: A European Perspective*, G. Bailey, ed. Cambridge University Press, Cambridge, pp.

Klein, R. G., 1987, Reconstructing how early people exploited animals: Problems and prospects, in: *The Evolution of Human Hunting*, M. H. Nitecki and D. V. Nitecki, eds., Plenum Press, New York, pp. 11-46.

Klein, R. G., 1994, Southern Africa before the Iron Age, in: *Integrative Paths to the Past: Paleoanthropological Advances in Honor of F. Clark Howell*, R. S. Corruccini and R. Ciochon, eds., Prentice Hall, Englewood Cliffs, N.J, pp. 471-519.

Klein, R. G. and K. Cruz-Uribe, 1996, Exploitation of large bovids and seals at Middle and Later Stone Age sties in South Africa, *Journal of Human Evolution* **31**:315-334.

Klein, R. G., 1999, *The Human Career: Human Biological and Cultural Origins*, University of Chicago Press, Chicago.

Klein, R. G., 2000, Archeology and the evolution of human behavior, *Evolutionary Anthropology* **9**:17-36.

Knight, C., C. Power and I. Watts, 1995, The human symbolic revolution: A Darwinian account, *Cambridge Archaeological Journal* **5**:1-27.

Koby, F., 1951, Grottes autrichiennes avec culte de l'ours? *Bulletin de la Société Préhistorique Française* **48**:8-9.

Kohler, T. A. and G. J. Gumerman, eds., 2000, *Dynamics in Human and Primate Societies: Agent-Based Modelling of Social and Spatial Processes*, Oxford University Press, Oxford.

Krings, M., A. Stone, R. W. Schmitz, H. Krainitzki, M. Stoneking et al., 1997, Neandertal DNA sequences and the origin of modern humans, *Cell* **90**:19-30.

Krings, M., H. Geisert, R. W. Schmitz, H. Krainitzki and S. Pääbo, 1999, DNA Sequence of the mitochondrial hypervariable region II from the Neandertal type specimen, *Proceedings of the National Academy of Science* **96**:5581-5585.

Kroeber, A. L., 1917, The superorganic, *American Anthropologist* **19**:163-213.

Kroeber, A. L., 1952, *The Nature of Culture*, University of Chicago Press, Chicago.

Kroeber, A. L. and C. Kluckhohn, 1952, *Culture; a Critical Review of Concepts and Definitions*, vol. 47 no. 1, Translated by Papers of the Peabody Museum of American Archaeology and Ethnology, Peabody Musem, Harvard University, Cambridge, MA.

Kuckenburg, M., 2001, *Als der Mensch zum Schöpfer wurde: An den Wurzeln der Kultur*, Klett-Cotta, Stuttgart.

Kurtén, B., 1976, *The Cave Bear Story: Life and Death of a Vanished Animal*, Columbia University Press, New York.

Laitman, J. T., R. C. Heimbuch and E. S. Crelin, 1978, Developmental change in a basicranial line and its relationship to the upper respiratory system in living primates, *American Journal of Anatomy* **152**:467-482.

Laitman, J. T., R. C. Heimbuch and E. S. Crelin, 1979, The basicranium of fossil hominids as an indicator of their upper respiratory systems, *American Journal of Physical Anthropology* **51**:15-34.

Laitman, J. T., J. S. Reidenberg, P. J. Gannon, B. Johansson, K. Landahl et al., 1990, The Kebara Hyoid: What can it tell us about the evolution of the hominid vocal tract? *American Journal of Physical Anthropology* **81**:254.

Laland, K. N. and W. Hoppitt, 2003, Do animals have culture? *Evolutionary Anthropology* **12**:150-159.

Lansing, J. S., 2000, Anti-chaos, common property, and the emergence of cooperation, in: *Dynamics in Human and Primate Societies: Agent-Based Modelling of Social and Spatial Processes*, T. A. Kohler and G. J. Gumerman, eds., Oxford University Press, Oxford, pp. 207-223.

Larson, J. E. and S. W. Herring, 1996, Movement of the epiglottis in mammals, *American Journal of Physical Anthropology* **100**:71-82.

Lartet, E. and H. Christy, 1865-1875, *Reliquiae Aquitanicae, Being Contributions to the Archaeology and Palaeontology of Périgord and the Adjoining Provinces of Southern France*, Williams & Norgate, London.

Lascu, C., F. Baciu, M. Gligan and S. Sarbu, 1996, A Mousterian cave bear worship site in Transylvania, Roumania, *Journal of Prehistoric Religion* **10**:17-30.

Latimer, B. and J. C. Ohman, 2001, Axial dysplasia in *Homo erectus*, *Journal of Human Evolution* **40**:A12.

Laville, H., 1988, Recent developments on the chronostratigraphy of the Paleolithic in the Perigord, in: *Upper Pleistocene Prehistory of Western Eurasia*, H. L. Dibble and A. Montet-White, eds., University Museum Press, Philadelphia, pp. 147-160.

Le Mort, F., 1987, Incisions volontaires sur un arrière-crâne de Néandertalien de Marillac (Charente), in: *Préhistoire de Poitou-Charentes: Problèmes Actuels*, Editions du C.T.H.S, Paris, pp. 151-156.

Le Mort, F., 1989a, Traces de décharnement sur les ossements néandertaliens de Combe-Grenal (Dordogne), *Bulletin de la Société Préhistorique Française* **86**:79-87.

Le Mort, F., 1989b, Le décharnement du cadavre chez les néandertaliens: quelques exemples, in: *L'Homme de Néandertal, Vol. 5, La Pensée*, O. Bar Yosef, ed. pp. 43-56.

Leakey, L. S. B., 1958, Recent discoveries at Olduvai Gorge, Tanzania, *Nature* **181**:1099-1103.

Leakey, M. D., 1971, *Olduvai Gorge, Volume 3. Excavations In Beds II & III, 1960-1963*, Cambridge University Press, Cambridge.

Lee, P. C., 1983, Caretaking of infants and mother-infant relationships, in: *Primate Social Relationships: An Integrated Approach*, R. A. Hinde, ed. Oxford University Press, Oxford, pp.

LeMay, M., 1975, The language capability of Neanderthal man, *American Journal of Physical Anthropology* **42**:9-14.

LeMay, M., 1976, Morphological cerebral asymmetries of modern man, fossil man, and nonhuman primates, in: *Origins and Evolution of Language and Speech*, H. B. Steklis, S. R. Harnad and J. Lancaster, eds., pp. 289-311.

LeMay, M., M. Billig and N. Geschwind, 1982, Asymmetries of the brains and skulls of nonhuman primates, in: *Primate Brain Evolution.*, E. Armstrong and D. Falk, eds., Plenum, New York, pp. 263-278.

LeMay, M., 1985, Asymmetries of the brains and skulls of nonhuman primates, in: *Cerebral Lateralization in Nonhuman Species*, S. D. Glick, ed. Academic Press, Orlando, pp. 233-245.

Leonardi, P., 1988, Art paléolithique mobilier et pariétal en Italie, *L'Anthropologie* **92**:139-202.

Leroi-Gourhan, A., 1947, La Grotte des Furtins (Commune de Berze-la-Ville, Saône-et-Loire), *Bulletin de la Société Préhistorique Française* **44**:43-55.

Leroi-Gourhan, A., 1959, Etude des restes humains fossiles provenants des grottes d'Arcy-sur-Cure, *Annales de Paléontologie* **44**:87-148.

Leroi-Gourhan, A., 1964a, *Le geste et la parole. La mémoire et les rythmes*, Sciences d'aujourd'hui, Albin Michel, Paris.

Leroi-Gourhan, A., 1964b, *Les Religions de la Préhistoire*, Presses Universitaires de France, Paris.

Leroi-Gourhan, A. and Arl. Leroi-Gourhan, 1965, Chronologie des grottes d'Arcy-sur-Cure (Yonne), *Gallia Préhistoire* **7**:1-64.

Leroi-Gourhan, Arl., 1975, The flowers found with Shanidar IV, A Neanderthal burial in Iraq, *Science* **190**:562-564.

Leroi-Gourhan, Arl., 1998, Shanidar et ses fleurs, *Paléorient* **24**:79-88.

Lévêque, F. and B. Vandermeersch, 1980, Découverte de restes humains dans un niveau castelperronien à Saint-Césaire (Charente-Maritime), *Comptes Rendus de l'Académie des Sciences de Paris* **Série D, 129**:187-189.

Levine, M. A., 1983, Mortality models and the interpretation of horse population structure, in: *Hunter-Gatherer Economy in Prehistory: A European Perspective*, G. Bailey, ed. Cambridge University Press, Cambridge, pp.

Lieberman, P., D. H. Klatt and W. H. Wilson, 1969, Vocal tract limitations on the vowel repertoires of rhesus monkey and other nonhuman primates, *Science* **164**:1184-1187.

Lieberman, P. and E. S. Crelin, 1971, On the speech of Neanderthal Man, *Linguistic Inquiry* **2**:203-222.

Lieberman, P., E. S. Crelin and D. H. Klatt, 1972, Phonetic ability and related anatomy of the newborn and adult human, Neanderthal man, and the chimpanzee, *American Anthropologist* **74**:287-307.

Lieberman, P., 1973, On the evolution of language: A unified view, *Cognition* **2**:59-94.

Lieberman, P. and E. S. Crelin, 1974, Speech and Neanderthal man: A reply to Carlisle and Siegel, *American Anthropologist* **76**:323-325.

Lieberman, P., 1976a, Structural harmony and Neanderthal speech: A reply to LeMay, *American Journal of Physical Anthropology* **43**:496.

Lieberman, P., 1976b, Interactive models for evolution: Neural mechanisms, anatomy, and behaviour, in: *Origins and Evolution of Language and Speech.*, S. R. Harnad and M. Stekelis, eds., pp. 660-672.

Lieberman, P., 1984, *The Biology and Evolution of Language*, Harvard University Press, Cambridge.

Lieberman, P., 1989, The origins of some aspects of human language and cognition, in: *The Human Revolution*, P. Mellars and C. B. Stringer, eds., Edinburgh University Press, Edinburgh, pp. 391-415.

Lieberman, P., J. T. Laitman and et al., 1989, Folk physiology and talking hyoids, *Nature* **342**:486-487.

Lieberman, P., 1991, *Uniquely Human: The Evolution of Speech, Thought, and Selfless Behavior*, Harvard University Press, Cambridge.

Lieberman, P., J. T. Laitman, J. S. Reidenberg and P. J. Gannon, 1992, The anatomy, physiology, acoustics and perception of speech: Essential elements in analysis of human speech, *Journal of Human Evolution* **23**:447-467.

Lieberman, P., 1993, On the Kebara KMH 2 hyoid and Neanderthal speech, *Current Anthropology* **34**:172-177.

Lieberman, P., 1994, Human language and human uniqueness, *Language and Communication* **14**:86-95.

Lieberman, P., 2001, On the neural bases of spoken language, in: *In the Mind's Eye : Multidisciplinary Approaches to the Evolution of Human Cognition*, A. Nowell, ed. International Monographs in Prehistory, Ann Arbor, pp. 172-186.

Lietava, J., 1992, Medicinal plants in a Middle Paleolithic grave at Shanidar IV? *Journal of Ethnopharmacology* **35**:263-266.

Lindly, J. M. and G. A. Clark, 1990, Symbolism and modern human origins, *Current Anthropology* **31**:233-261.

Lombard, M., 2005, Evidence of hunting and hafting during the Middle Stone Age at Sibidu Cave, KwaZulu-Natal, South Africa: A multianalytical approach, *Journal of Human Evolution* **48**:279-300.

Lowie, R. H., 1954, *Indians of the Plains*, American Museum of Natural History, Anthropological Handbook no. 1, McGraw-Hill, New York.

Lyman, R. L., 1987, On the analysis of vertebrate mortality profiles: Sample size, mortality type, and hunting pressure, *American Antiquity* **52**:125-142.

MacLarnon, A. M., 1993, The vertebral canal, in: *The Nariokotome Homo erectus skeleton*, A. Walker and R. E. F. Leakey, eds., Harvard University Press, Cambridge, pp. 359-390.

MacLarnon, A. M. and G. P. Hewitt, 1999, The evolution of human speech: The role of enhanced breathing control, *American Journal of Physical Anthropology* **109**:341-363

Mainzer, K., 1997, *Thinking in Complexity: The Complex Dynamics of Matter, Mind, and Mankind*, Springer, Berlin.

Malinowski, B., 1944, *A Scientific Theory of Culture and Other Essays*, University of North Carolina Press, Chapel Hill.

Malinowski, B., 1954, *Magic, Science, and Religion and Other Essays*, Doubleday, Garden City, NY.

Mandelbaum, D. G., ed. 1968, *Selected Writings of Edward Sapir in Language, Culture, and Personality*, University of California Press, Berkeley.

Mandl, I., 1961, Collagenases and elastases, *Advances in Enzymology and Related Subjects of Biochemistry* **23**:164-264.

Mania, D. and T. Weber, 1986, *Bilzingsleben III.*, VEB Deuscher Verlag der Wissenschaften, Berlin.

Mania, D. and U. Mania, 1988, Deliberate Engravings on Bone Artifacts of *Homo Erectus, Rock Art Research* **5**:91-107.

Mania, D., ed. 1990, *Neumark – Gröbern. Beiträge zur Jagd des mittlepaläolithischen Menschen*, Deutshcer Verlag der Wissenschaften, Berlin.

Mann, A. and E. Trinkaus, 1973, Neandertal and Neandertal-like fossils from the Upper Pleistocene, *Yearbook of Physical Anthropology* **17**:169-193.

Maringer, J., 1960, *The Gods of Prehistoric Men*, Knopf, New York.

Marks, A. E., H. J. Hietala and J. K. Williams, 2001, Tool standardization in the Middle and Upper Palaeolithic: A closer look, *Cambridge Archaeological Journal* **11**:17-44.

Marler, P. and M. Tamura, 1964, Culturally transmitted patterns of vocal behavior in sparrows, *Science* **146**:1483-1486.

Marquet, J.-C. and M. Lorblanchet, 2003, A Neanderthal face? The proto-figurine from La Roche-Cotard, Langeais (Indre-et-Loire, France), *Antiquity* **77**:661-670.

Marshack, A. A., 1976, Some implications of the Paleolithic symbolic evidence for the origins of language, *Current Anthropology* **17**:274-282.

Marshack, A. A., 1985, Theoretical concepts that lead to new analytic methods, modes of inquiry and classes of data, *Rock Art Research* **2**:95-111.

Marshack, A. A., 1988, The Neanderthals and the human capacity for symbolic thought : Cognitive and problem-solving aspects of mousterian symbolism, in: *L'Homme de Néandertal. Actes du Colloque International de Liège (4-7 décembre 1986). Vol. 5: La Pensée*, M. Otte, ed. ERAUL, Liège, pp. 57-91, 12 fig.

Marshack, A. A., 1989, Neanderthals and the human capacity for symbolic thought: Cognitive and problem-solving aspects of Mousterian symbol, in: *L'Homme de Néandertal. Actes du Colloque International de Liège (4-7 décembre 1986). Vol. 5: La Pensée*, M. Otte, ed. ERAUL, Liège, pp. 57-92.

Marshack, A. A., 1990, Early hominid symbol and evolution of the human capacity, in: *The Emergence of Modern Humans: An Archaeological Perspective*, P. Mellars, ed. Cornell University Press, Ithica, pp. 457-498.

Marshack, A. A., 1991, A reply to Davidson on Mania & Mania, *Rock Art Research* **8**:47-58.

Marshack, A. A., 1997, The Berekhat Ram figurine: A late Acheulian carving from the Middle East, *Antiquity* **71**:327-337.

Martin, H., 1906, Présentation d'ossements de renne portant des traces des lésion d'origine humaine et animale, *Bulletin de la Société Préhistorique Française* **3**:385-391.

Martin, H., 1907-1910, *Recherches sur l'évolution du Moustérien dans le gisement de La Quina (Charente). Vol 1: Industrie osseuse*, Schleicher Frères, Paris.

Martin, H., 1909, *Recherches sur l'évolution du Moustérien dans le gisement de La Quina (Charente), vol. 2*, Schleicher Frères, Paris.

Marx, K., 1884 [1970], *Marx's Critique of Hegel's Philosophy of Right (1843)*, Oxford University Press, Oxford.

Mason, R. J., 1988, *Cave of Hearths, Makapansgat, Transvaal*, University of the Witwatersrand Archaeological Research Unit, Occasional Paper 21, Johannesburg.

Maynard Smith, J., 1964, Group selection and kin selection, *Nature* **201**:1145-1147.

Maynard Smith, J., 1976, Group selection, *Quarterly Review of Biology* **51**:277-283.

Maynard Smith, J., 1999, Conflict and cooperation in human societies, in: *Levels of Selection in Evolution.*, L. Keller, ed. Princeton University Press, Princeton, pp. 197-208.

McBrearty, S. and A. S. Brooks, 2000, The revolution that wasn't: A new interpretation of the origin of modern human behavior, *Journal of Human Evolution* **39**:453-563.

McBurney, C. B. M. and P. Callow, 1971, The Cambridge excavations a La Cotte de St. Brelade, Jersey - A preliminary report, *Proceedings of the Prehistoric Society* **37**:167-207.

McGrew, W. C., C. E. G. Tutin and P. Baldwin, 1979, Chimpanzees, tools, and termites: Cross-cultural comparisons of Senegal, Tanzania, and Rio Muni, *Man* **14**:185-215.

McGrew, W. C., L. F. Marchant, T. Nishida, J. Goodall and J. Itani, eds., 1996, *Great Ape Societies*, Cambridge University Press, Cambridge.

McGrew, W. C., 1998, Culture in non-human primates? *Annual Review of Anthropology* **27**:301-328.

McNabb, J., F. Binyon and L. Hazelwood, 2004, The large cutting tools from the South African Acheulean and the question of social traditions, *Current Anthropology* **45**:653-657.

McPherron, S. P., 1994, *A Reduction Model for Variability in Acheulean Biface Morphology*. PhD dissertation, University of Pennsylvania.

McPherron, S. P., 1995, A re-examination of the British biface data, *Lithics* **16**:47-63.

McPherron, S. P., 2000, Handaxes as a measure of the mental capabilities of early hominids, *Journal of Archaeological Science* **27**:655-664.

Mech, L. D., 1970, *The Wolf: The Ecology and Behavior of an Endangered Species*, Natural History Press, Garden City, NY.

Meignen, L. and O. Bar-Yosef, 1992, Middle Paleolithic variability in Kebara Cave, Mount Carmel, Israel, in: *The Evolution and Dispersal of Modern Humans in Asia*, T. Akazawa, K. Aoki and T. Kimura, eds., Hokusen-Sha, Tokyo, pp. 129-148.

Mellars, P., 1973, The character of the Middle-Upper Paleolithic transition in Southwest France, in: *The Explanation of Culture Change: Models in Prehistory*, C. Renfrew, ed. University of Pittsburgh Press, Pittsburgh, pp. 255-276.

Mellars, P., 1988, The chronology of the South-West French Mousterian: A review of the current debate, in: *L'Homme de Neandertal, Vol. 4, La Technique*, L. Binford and J.-P. Rigaud, eds., Université de Liège, Liège, pp. 97-120.

Mellars, P., 1989, Major issues in the emergence of modern humans, *Current Anthropology* **30**:349-385.

Mellars, P., 1991, Cognitive Changes and the Emergence of Modern Humans in Europe, *Cambridge Archaeological Journal* **1**:63-76.

Mellars, P., 1996a, *The Neanderthal Legacy: An Archaeological Perspective from Western Europe*, Princeton University Press, Princeton.

Mellars, P., 1996b, Symbolism, language and the Neanderthal mind, in: *Modelling the Early Human Mind*, P. Mellars and K. R. Gibson, eds., Cambridge, pp. 15-32.

Mellars, P., 1999, The Neanderthal problem continued, *Current Anthropology* **40**:341-364.

Mellars, P., 2000a, Châtelperronian chronology and the case for Neanderthal/modern human 'acculturation' in Western Europe, in: *Neanderthals on the Edge. Papers from a Conference Marking the 150th Anniversary of the Forbes' Quarry Discovery, Gibraltar*, C. B. Stringer, R. N. E. Barton and J. C. Finlayson, eds., Oxbow Books, Oxford, pp. 33-39.

Mellars, P., 2000b, The archaeological records of the Neanderthal-modern human transition in France, in: *The Geography of Neandertals and Modern Humans in Europe and the Greater Mediterranean.*, O. Bar Yosef and D. Pilbeam, eds., Cambridge, MA, pp. 35-48.

Mellars, P., 2005, The impossible coincidence. A single-species model for the origins of modern human behavior in Europe, *Evolutionary Anthropology* **14**:12-27.

Mercier, N. and H. Valladas, 1998, Datation, *Gallia Préhistoire* **40**:70-71.

Mertl-Millhollen, A. S., 2000, Tradition in *Lemur catta* behavior at Berenty Reserve, Madagascar, *International Journal of Primatology* **21**:287-297.

Meyer, M. R., 2003, Vertebrate and language ability in early hominids, Abstracts, Paleoanthropological   Society   2003   Meeting. (http://www.pennpress.org/pa/paleoabstracts2003/paleoabs.-pdf).

Mihata, K., 1997, The persistence of "emergence", in: *Chaos, Complexity and Sociology: Myths Models and Theories.*, R. A. Eve, S. Horsfall and M. E. Lee, eds., Sage, 30-38, pp.

Miles, H. L. W., 1990, The cognitive foundations for reference in a signing orangutan, in: *"Language" and Intelligence in Monkeys and Apes*, S. T. Parker and K. R. Gibson, eds., Cambridge University Press, Cambridge, pp. 511-539.

Miller, G. H., P. B. Beaumont, H. J. Deacon, A. Brooks, P. E. Hare et al., 1999, Earliest modern humans in South Africa dated by isoleucine epimerization in ostrich eggshell, *Quarternary Science Reviews* **18**:1537-1548.

Mithen, S., 1996a, Domain-specific intelligence and the Neanderthal mind, in: *Modelling the Early Human Mind*, P. Mellars and K. R. Gibson, eds., Cambridge, pp. 217-229.

Mithen, S., 1996b, *Prehistory of the Mind: The Cognitive Origins of Art, Religion, and Science*, Thames and Hudson, New York.

Monnier, G. F., 2000, *A Re-Evaluation of the Archaeological Evidence for a Lower/Middle Paleolithic Division in Western Europe.* PhD. Dissertation, University of Pennsylvania.

Montagu, A., 1976, Toolmaking, hunting, and the origin of language, in: *Origins and Evolution of Language and Speech.*, H. B. Steklis, S. R. Hamad and J. Lancaster, eds., New York, pp. 266-274.

Mooney, J., 1896, *The Ghost-Dance Religion and the Sioux Outbreak of 1890*, Bureau of American Ethnography, Annual Report, Part 2, Washington.

Morel, J., 1974, La station éponyme de l'Oued Djebbana à Bir-el-Ater (est algérien). Contribution à la connaissance de son industrie et de sa faune, *L'Anthropologie* **78**:53-80.

Morris, D. H., 1974, Neanderthal speech, *Linguistic Inquiry* **5**:144-150.

Mottl, M., 1951, Die Repolust-Höhle bei Peggau (Steiermark) und ihre eiszeitlichen Bewohner, *Archaeologia Austriaca* **8**:1-78.

Movius, H. L., 1950, A wooden spear of third interglacial age from lower Saxony, *Southwestern Journal of Anthropology* **6**:139-142.

Murdoch, J., 1892/1988, *Ethnological Results of the Point Barrow Expedtion*, Smithsonian Institution Press, Washington, D.C.

Myers Thompson, J. A., 1994, Cultural diversity in the behavior of *Pan*, in: *Hominid Culture in Primate Perspective*, D. Quiatt and J. Itani, eds., University Press of Colorado, Niwor, CO, pp. 95-115.

Nellemann, G., 1969/70, Caribou hunting in west Greenland, *Folk* **11-12**:133-153.

Netting, R. M., 1974, The system nobody knows: Village irrigation in the Swiss Alps, in: *Irrigation's Impact on Society*, M. Gibson and T. E. Downing, eds., University of Arizona Press, Tucson, pp. 67-75.

Nicholson, N., 1987, Infants, mothers, and other females, in: *Primate Societies*, B. B. Smuts, D. L. Cheney, R. M. Seyfarth, R. W. Wrangham and T. T. Struhsaker, eds., University of Chicago Press, Chicago, pp.

Nicolis, G. and I. Prirogine, 1977, *Self-Organization in Nonequilibrium Systems: From Disspative Structures to Order through Fluctuations*, Wiley, New York.

Nicolis, G. and I. Prigogine, 1989, *Exploring Complexity : An Introduction*, W.H. Freeman, New York.

Nishida, T., 1986, Local Traditions and Cultural Transmission, in: *Primate Societies.*, B. Smuts, D. L. Cheney, R. M. Seyfarth, R. W. Wrangham and T. T. Struhsaker, eds., University of Chicago Press, Chicago, pp. 462-474.

Nishida, T., J. C. Mitani and D. P. Watts, 2004, Variable Grooming Behaviours in Wild Chimpanzees, *Folia Primatologica* **75**:31-36.

Nishimura, T., A. Mikami, J. Suzuki and T. Matsuzawa, 2003, Descent of the larynx in chimpanzee infants, *Proceedings of the National Academy of Sciences* **100**:6930-6933.

Noble, W. and I. Davidson, 1991, The evolutionary emergence of modern human behaviour: Language and its archaeology, *Man* **26**:223-253.

Noble, W. and I. Davidson, 1996, *Human Evolution, Language and Mind: A Psychological and Archaeological Inquiry*, Cambridge University Press, Cambridge.

Noble, W. and I. Davidson, 2001, Discovering the symbolic potential of communicative signs: The origins of speaking a language, in: *In the Mind's Eye : Multidisciplinary Approaches to the Evolution of Human Cognition*, A. Nowell, ed. International Monographs in Prehistory, Ann Arbor, pp. 187-200.

Nordborg, M., 1998, On the probability of Neanderthal ancestry, *American Journal of Human Genetics* **63**:1237-1240.

Nowell, A., 2000, *The Archaeology of Mind: Standardization and Symmetry in Lithics and Their Implications for the Study of the Evolution of the Human Mind.* PhD. Dissertation, University of Pennsylvania.

Nowell, A. and P. G. Chase, 2002, Is a cave bear bone from Divje Babe, Slovenia, a Neanderthal flute? in: *The Archaeology of Early Sound: Origin and Organization.2nd Symposium of the Study Group on Music Archaelogy*, E. Hickmann and R. Eichmann, eds., pp. 69-81.

Oakley, K., P. Andrews, L. H. Keeley and J. D. Clark, 1977, A reappraisal of the Clacton spearpoint, *Proceedings of the Prehistoric Society* **43**:13-30.

Oakley, K. P., 1971, Fossils collected by the earlier palaeolithic men, in: *Mélanges de préhistoire, d'archéocivilization et d'ethnologie offerts à André Varagnac*, Serpen, Paris, pp. 581-584.

Oakley, K. P., 1973, Fossil shell observed by Acheulian man, *Antiquity* **47**:59-60, plate XIa.

Oakley, K. P., 1981, The emergence of higher thought 3.0-0.2 Ma B.P, *Philosophical Transactions of the Royal Society, London* **292B**:205-211.

O'Meara, T., 1997, Causation and the struggle for a science of culture, *Current Anthropology* **38**:399.

Otte, M., G.-M. Cordy and D. Mangon, 1985, Dents incisées du Paléolithique moyen, *Cahiers de Préhistoire liégeoise. I. Rapport des activités* 80-84.

Pales, L., 1954, Les empreintes des pieds humains de la "Tana della Basura" (Toirano), *Rivista di Studi Liguri* **20**:5-12.

Parker, S. T. and K. R. Gibson, 1979, A developmental model for the evolution of language and intelligence in early hominids, *Behavioral and Brain Sciences* **2**:367-408.

Patterson, F., 1978, Linguistic capabilities of a lowland gorilla, in: *Sign Language and Language Acquisition in Man and Ape*, F. C. C. Peng, ed. Westview Press, Boulder, CO, pp. 161-201.

Peirce, C. S., 1932 [1960], The icon, index, and symbol, in: *Collected Papers of Charles Sanders Pierce, vol. II*, C. Hartshorne and P. Weiss, eds., Harvard University Press, Cambridge, pp. 156-173.

Pelcin, A., 1994, A geological explanation for the Berekhat Ram figurine, *Current Anthropology* **35**:674-675.

Pepper, J. W. and B. B. Smuts, 2000, The evolution of cooperation in an ecological context: An agent-based model, in: *Dynamics in Human and Primate Societies: Agent-*

Based Modelling of Social and Spatial Processes, T. A. Kohler and G. J. Gumerman, eds., Oxford University Press, Oxford, pp. 45-76.

Perry, S., M. Baker, L. Fedigan, K. Jack, K. C. Mackinnon et al., 2003, Social conventions in wild white-faced capuchin monkeys: Evidence for traditions in a neotropical primate, *Current Anthropology* **44**:241-268.

Philibert, S., 1994, L'ocre et le traitement des peaux: Révision d'une conception traditionnelle par l'analyse fonctionnelle des grattoirs ocrés de la Balma Margineda (Andorre), *L'Anthropologie* **98**:447-453.

Piaget, J., 1952, *The Origins of Intelligence in Children*, W.W. Norton and Company, New York.

Piaget, J., 1954, *The Construction of Reality in the Child*, Basic Books, New York.

Piaget, J., 1955, *Language and Thought of the Child*, World Publishing Co, New York.

Pinker, S., 1994, *The Language Instinct*, William Morrow, New York.

Poole, J. and D. G. Lander, 1971, The pigeon's concept of pigeon, *Psychonomic Science* **25**:157-158.

Potts, R., 1984, Home bases and early hominids, *American Scientist* **72**:338-347.

Potts, R., 1987, Reconstructions of early hominid socioecology: A critique of primate models, in: *The Evolution of Human Behavior: Primate Models.*, W. G. Kinzey, ed. State University of New York Press, Albany, NY, pp. 28-47.

Premack, D. and A. J. Premack, 1983, *The Mind of an Ape*, W.W. Norton and Company, New York.

Quiatt, D. and V. A. Reynolds, 1993, *Primate Behaviour: Information, Social Knowledge, and the Evolution of Culture*, Cambridge University Press, Cambridge.

Quiatt, D. and J. Itani, 1994, *Hominid Culture in Primate Perspective*, University Press of Colorado, Niwor, CO.

Radcliffe-Brown, A. R., 1935/1952, On the concept of function in social science, in: *Structure and Function in Primitive Society*, A. R. Radcliffe-Brown, ed. Free Press, Glencoe, IL, pp. 179-187.

Rappaport, R. A., 1999, *Ritual and Religion in the Making of Humanity*, Cambridge Studies in Social and Cultural Anthropology 10, Cambridge University Press, Cambridge.

Raynal, J.-P. and R. Seguy, 1986, Os incisé acheuléen de Sainte-Anne 1 (Polignac, Haute-Loire), *Revue archéologique du centre* **25**:79-81.

Reader, S. M. and K. N. Laland, 2002, Social intelligence, innovation, and enhanced brain size in primates, *Proceedings of the National Academy of Sciences* **99**:4436-4441.

Reeve, H. K. and L. Keller, 1999, Levels of selection: Burying the units-of-selection debate and unearthing the crucial new issues, in: *Levels of Selection in Evolution*, L. Keller, ed. Princeton University Press, Princeton, pp. 3-14.

Relethford, J. H., 2001, Absence of regional affinities of Neandertal DNA with living humans does not reject multiregional evolution, *American Journal of Physical Anthropology* **115**:95-98.

Renault-Miskovsky, J., 1998, Palynologie, *Gallia Préhistoire* **40**:65-68.

Reynolds, P. J., 1993, The complementation theory of language and tool use, in: *Tools, Language and Cognition in Human Evolution.*, K. R. Gibson and T. Ingold, eds., Cambridge University Press, Cambridge, pp. 407-428.

Richerson, P. J. and R. R. Boyd, 1998, The evolution of human ultra-sociality, in: *Indoctrinability, Ideology, and Warfare*, I. Eibl-Eibesfeldt and F. K. Salter, eds., Berghahn, New York, pp.

Richerson, P. J. and R. R. Boyd, 1999, Complex societies - the origins of a crude superorganism, *Human Nature* **10**:253-289.

Rightmire, G. P., 1981, Later Pleistocene hominids of eastern and southern Africa, *Anthropologie (Brno)* **19**:15-26.

Rightmire, G. P., 1984, The fossil evidence for hominid evolution in southern Africa, in: *Southern African Prehistory and Paleoenvironments*, R. G. Klein, ed. A.A. Balkema, Rotterdam, pp. 147-168.

Rightmire, G. P., 1998, Human evolution in the Middle Pleistocene: The role of *Homo heidelbergensis*, *Evolutionary Anthropology* **6**:218-227.

Rightmire, G. P., 2004, Brain size and encephalization in early to Mid-Pleistocene *Homo*, *American Journal of Physical Anthropology* **124**:109-123.

Rindos, D., 1985, Darwinian selection, symbolic variation, and the evolution of culture, *Current Anthropology* **26**:65-88.

Rindos, D., 1989, Undirected variation and the Darwinian explanation of cultural change, *Archaeological Method and Theory* **1**:1-45.

Ristau, C. A. and D. Robbins, 1982, Language in the great apes: A critical review, *Advances in the Study of Behavior* **12**:141-255.

Roberts, A. J., R. N. E. Barton and J. Evans, 1998, Early Mesolithic mastic: Radiocarbon dating and analysis of organic residues from Thatcham III, Star Carr and Lackford Heath, in: *Stone Age Archaeology: Essays in honour of John Wymer*, N. Ashton, F. Healy and P. B. Pettitt, eds., Oxbow Monograph 102, Oxford, pp. 185-192.

Rodseth, L., 1998, Distributive models of culture: A Sapiraian alternative to essentialism, *American Anthropologist* **100**:55-69.

Rose, N., 1998, Controversies in meme theory, Journal of Memetics 2:[http://jom-emit.cfpm.org/].

Ross, C. F. and M. Henneberg, 1995, Basicranial flexion, relative brain size, and facial kyphosis in *Homo sapiens* and some fossil hominids, *American Journal of Physical Anthropology* **98**:575-593.

Rumbaugh, D. M., ed. 1977, *Language Learning by a Chimpanzee: The LANA Project*, Academic Press, New York.

Russell, M., 1987a, Mortuary practices at the Krapina Neanderthal site, *American Journal of Physical Anthropology* **72**:381-397.

Russell, M., 1987b, Bone breakage in the Krapina hominid collection, *American Journal of Physical Anthropology* **72**:373-379.

Sackett, J. R., 1973, Style, function, and artifact variability in Palaeolithic assemblages, in: *The Explanation of Culture Change: Models in Prehistory*, C. Renfrew, ed. University of Pittsburgh Press, Pittsburgh, pp. 317-325.

Sackett, J. R., 1982, Approaches to style in lithic archaeology, *Journal of Anthropological Archaeology* **1**:59-112.

Sackett, J. R., 1985, Style, ethnicity, and stone tools, in: *Status, Structure and Stratification: Current Archaeological Reconstructions*, M. Thompson, M. T. Garcia and F. J. Kense, eds., University of Calgary, Calgary, pp. 277-282.

Sackett, J. R., 1986, Isochrestism and style: A clarification, *Journal of Anthropological Archaeology* **5**:266-277.

Sahlins, M., 1999, Two or three things that I know about culture, *Journal of the Royal Anthropological Institute* **5**:399-421.

Sapir, E., 1938, Why cultural anthropology needs the psychiatrist, *Psychiatry* **1**:7-12.

Savage-Rumbaugh, E. S., 1986, *Ape Language*, Columbia University Press, New York.

Savage-Rumbaugh, E. S., K. M. McDonald, R. Sevcik, W. D. Hopkins and E. Rubert, 1986, Spontaneous symbol acquisition and communicative use by pygmy chimpanzees (*Pan paniscus*), *Journal of Experimental Psychology: General* **115**:211-235.

Savage-Rumbaugh, E. S., 1987, Communication, symbolic communication, and language: Reply to Seidenberg and Pettito, *Journal of Experimental Psychology: General* **116**:288-292.

Savage-Rumbaugh, E. S., S. L. Williams, T. Furuichi and T. Kano, 1996, Language perceived: *Paniscus* branches out, in: *Great Ape Societies*, W. C. McGrew, L. F. Marchant and T. Nishada, eds., Cambridge University Press, Cambridge, pp. 173-184.

Savage-Rumbaugh, E. S., S. G. Shankar and T. J. Taylor, 1998, *Apes, Language and the Human Mind*, Oxford University Press, Oxford.

Schaller, G. B., 1972, *The Serengeti Lion: A Study of Predator-Prey Relations*, Wildlife Behavior and Ecology Series, University of Chicago Press, Chicago.

Schepartz, L. A., 1993, Language and modern human origins, *Yearbook of Physical Anthropology* **36**:91-126.

Schoenemann, T. P., 1997, *An MRI Study of the Relationship Between Human Neuroanatomy and Behavioral Ability*, PdD Dissertaion. University of California, Berkeley.

Scholz, M., L. Bachmann, G. J. Nicholson, J. Bachmann, S. Hengst et al., 2000, How different are the genomes of Neanderthals and anatomically modern man? A DNA hybridization approach, in: *Neanderthals and Modern Humans: Discussing the Transition*, J. Orscheidt and G.-C. Weniger, eds., Neanderthal Museum, Mettmann, pp. 31522.

Schwarcz, H. P. and I. Skoflek, 1982, New dates from the Tata, Hungary, archaeological site, *Nature* **295**:590-591.

Schwartz, J. H. and I. Tattersall, 2002, *The Human Fossil Record, Vol. 1. Terminology and Craniodental Morphology of Genus Homo (Europe)*, Wiley-Liss, New York.

Schwartz, T., 1978, Where is culture? Personality as the distributive locus of culture, in: *The Making of Psychological Anthropology*, G. D. Spindler, ed. University of California Press, Berkeley, pp. 419-441.

Scott, K., 1980, Two hunting episodes of middle palaeolithic age at La Cotte de Saint-Brelade, Jersey (Channel Islands), *World Archaeology* **12**:137-152.

Seidenberg, M. S. and L. A. Petitto, 1987, Communication, symbolic communication, and language: Comment on Savage-Rumbaugh, McDonald, Sevcik, Hopkins, and Rupert (1986), *Journal of Experimental Psychology: General* **116**:279-287.

Semendeferi, K., 2001, Before or after the split? Hominid neural specializations, in: *In the Mind's Eye : Multidisciplinary Approaches to the Evolution of Human Cognition*, A. Nowell, ed. International Monographs in Prehistory, Ann Arbor, pp.

Sevick, R. A. and E. S. Savage-Rumbaugh, 1994, Language comprehension and use by great apes, *Language and Communication* **14**:37-58.

Shea, J. J., 1988, Spear points from the Middle Paleolithic of the Levant, *Journal of Field Archaeology* **15**:441-450.

Shea, J. J., 1997, Middle Paleolithic spear point technology, in: *Projectile Technology*, H. Knecht, ed. Plenum Press, New York, pp. 79-106.

Shennon, S., 2003, *Genes, Memes and Human History: Darwinian Archaeology and Cultural Evolution*, Thames and Hudson, New York.

Sigg, H. and A. Stolba, 1981, Home range and daily march in a Hamadryas baboon troop, *Folia Primatologica* **36**:40-75.

Simon, H., 1990, A mechanism for social selection and successful altruism, *Science* **250**:1665-1668.

Singer, R. and J. Wymer, 1982, *The Middle Stone Age at Klasies River Mouth in South Africa*, University of Chicago Press, Chicago.

Smirnov, Y., 1989, Intentional human burial: Middle Paleolithic (last glaciation) beginnings, *Journal of World Prehistory* **3**:199-233.

Smuts, B., D. L. Cheney, R. M. Seyfarth, R. W. Wrangham and T. T. Struhsaker, 1986, *Primate Societies*, University of Chicago Press, Chicago.

Smuts, B., 1999, Multilevel selection, cooperation, and altruism: Reflections on *Unto Others: The Evolution and Psychology of Unselfish Behavior*, *Human Nature* **10**:311-327.

Snowdon, C. T., 1990, Language capacities of nonhuman animals, *Yearbook of Physical Anthropology* **33**:215-243.

Sober, E. and D. S. Wilson, 1998, *Unto Others: The Evolution and Psychology of Unselfish Behavior*, Harvard University Press, Cambridge, MA.

Solecki, R. S., 1975, Shanidar IV, a Neanderthal Flower burial in Northern Iraq, *Science* **190**:880-881.

Solecki, R. S., 1977, The implications of the Shanidar Cave Neanderthal flower burial, in: *Anthropology and the Climate of Opinion*, S. Freed, ed. New York, pp. 114-124.

Solecki, R. S., 1982, A ritual Middle Palaeolithic deer burial at Nahr Ibrahim Cave, Lebanon, in: *Archéologie au Levant: Recueil à la mémoire de Roger Saidah*, Maison de l'Orient, Lyon, pp. 47-56.

Soltis, J., R. R. Boyd and P. J. Richerson, 1995, Can group-functional behaviors evolve by cultural group selection? An empirical test, *Current Anthropology* **36**:473-494.

Sommer, J. D., 1999, The Shanidar IV "flower burial": A re-evaluation of Neanderthal burial ritual, *Cambridge Archaeological Journal* **9**:127-129.

Soressi, M. and C. S. Henshilwood, 2004, Blombos Cave, South Africa: Stone artifacts from the Middle Stone Age levels, *Paper presented at the Paleoanthropological Society Meeting, Montreal, 2004*.

Speth, J. D., 1997, Communal bison hunting in western North America: Background for the study of Paleolithic bison hunting in Europe, in: *L'Alimentation des hommes du Paléolithique: Approche pluridisciplinaire*, M. Patou-Mathis, ed. Université de Liège, Liège, pp. 23-58.

Spier, L., 1935, *The Prophet Dance of the Northwest and Its Derivatives: The Source of the Ghost Dance*, General Series in Anthropology, 1, George Banta Publishing Company, Menasha, WI.

Spitzer, M., 1999, *The Mind within the Net: Models of Learning, Thinking, and Acting*, MIT Press, Cambridge, MA.

Stefánsson, V., 1913/1962, *My Life with the Eskimo*, Collier Books, New York.

Stepanchuk, V. N., 1993, Prolom II, a Middle Palaeolithic cave site in the eastern Crimea with non-utilitarian bone artefacts, *Proceedings of the Prehistoric Society* **59**:17-37.

Stiles, D. N., 1979, Paleolithic culture and culture change: Experiment in theory and method, *Current Anthropology* **20**:1-21.

Stiner, M. C., 1991, The faunal remains from Grotta Guattari: A taphonomic perspective., *Current Anthropology* **32**:103-.

Stout, D., N. Toth, K. Schick, J. Stout and G. Hutchins, 2000, Stone tool-making and brain activation: Positron emission tomography (PET) studies, *Journal of Archaeological Science* **27**:1215-1223.

Strait, D. S., 1999, The scaling of basicranial flexion and length, *Journal of Human Evolution* **37**:701-719.

Stringer, C. B., 1981, The dating of European Middle Pleistocene hominids and the existence of *Homo erectus* in Europe, *Anthropologie (Brno)* **19**:3-14.

Stringer, C. B. and C. Gamble, 1993, *In Search of the Neanderthals: Solving the Puzzle of Human Origins*, Thames and Hudson, London.

Taborin, Y., 1988, Les prémices de l'expression symbolique, in: *De Néandertal à Cro-Magnon*, C. Farizy, ed. Musée de Préhistoire de l'Ile-de-France, Nemours, pp. 73-76.

Taborin, Y., 1990, Les prémices de la parure, in: *Paléolithique moyen récent et Paléolithique supérieur ancien en Europe: Colloque international de Nemours, 9-11 mai 1988*, C. Farizy, ed. Paris, pp. 335-344.

Tappen, N. C., 1987, Circum-mortem damage to some ancient African hominid crania: A taphonomic and evolutionary essay, *African Archaeological Review* **5**:39-47.

Terrace, H. S., 1979, Is problem-solving language? *Journal of the Explanatory Analysis of Behavior* **31**:161-175.

Terrace, H. S., L. A. Petitto and T. G. Bever, 1979, Can an ape create a sentence? *Science* **206**:891-900.

Texier, J.-P., J. Huxtable, E. J. Rhodes, D. Miallier and M. Ousmoi, 1988, Nouvelles données sur la situation chronologique de l'Atérien du Maroc et leurs implications, *Comptes Rendus de l'Académie des Sciences (Paris) Série II* **307**:827-832.

Thévenin, A., 1976, Les civilisations du Paléolithique inférieur en Alsace, in: *La Préhistoire Française. Tome I. Les Civilizations Paléolithique et Mésolithique de la France*, H. de Lumley, ed. CNRS, Paris, pp. 984-986.

Thieme, H. and S. Veil, 1985, Neue Untersuchungen zum eemzeitlichen Elefanten-Jagdplatz Lehringen, Landkr. Verden, *Die Kunde n.F.* **36**:11-58.

Thieme, H., 1997, Lower Palaeolithic hunting spears from Germany, *Nature* **385**:807-810.

Thompson, J., J. Bower, E. Fischer, A. Mabulla, C. Marean et al., 2004, Loiyangalani: Behavioral and taphonomic aspects of a Middle Stone Age site in the Serengeti Plain, Tanzania, *Paper presented at the Paleoanthropological Society Meeting, Montreal, 2004.*

Tillet, T., 1978, Présence de pendeloques en milieu atérien au Niger Oriental., *Bulletin de la Société Préhistorique Française* **75**:273-275.

Tillier, A.-M., B. Arensburg, Y. Rak and B. Vandermeersch, 1988, Les sépultures Néanderthaliennes du Proche-Orient: État de la question, *Paléorient* **14**:130-136.

Tillier, A.-M., B. Arensburg, B. Vandermeersch and Y. Rak, 1991, L'apport de Kébara à la palethnologie funéraire des Néanderthaliens du Proche-Orient, in: *Le Squelette Moustérien de Kébara 2*, O. Bar Yosef and V. Bernard, eds., C.N.R.S, Paris, pp. 89-96.

Tillier, A.-M., 1995, Paléoanthropologie et pratiques funéraires au Levant méditerranéen durant le Paléolithique moyen: Le cas des sujets non-adultes, *Paléorient* **21**:63-76.

Tixier, J., 1960, Les industries lithiques d'Ain Fritissa (Maroc oriental), *Bulletin d'archéologie marocaine* **3**:107-244.

Tobias, P. V., 1987, The brain of *Homo habilis:* A new level of organization in cerebral evolution, *Journal of Human Evolution* **16**:741-761.

Tobias, P. V., 1991, The emergence of spoken language in hominid evolution, in: *Cultural Beginnings: Approaches to Understanding Early Hominid Life-Ways in the African Savannah.*, J. D. Clark, ed. Rudolf Habelt, Bonn, pp. 67-78.

Tomasello, M. and J. Call, 1994 Social cognition of monkeys and apes, *Yearbook of Physical Anthropology* **37**:273-305.

Tomasello, M. and J. Call, 1997, *Primate Cognition*, Oxford University Press, New York.

Tomasello, M., 1999, The human adaptation for culture, *Annual Review of Anthropology* **28**:509-528.

Tomasello, M., M. Carpenter, J. Call, T. Behne and H. Moll, in press, Understanding and sharing intentions: The origins of cultural cognition, *Behavioral and Brain Sciences.*

Toth, N. and T. D. White, 1990-1991, Assessing the ritual cannibalism hypothesis at Grotta Guattari, in: *The Fossil Man of Monte Circeo: 50 Years of Studeis on the Neandertals in Latium*, A. Bietti and G. Manzi, eds., pp. 213-222.

Tracy, L., 1996, Genes, memes, templates and replicators, *Behavioral Science* **41**:205-214.

Trinkaus, E., 1985, Cannibalism and burial at Krapina, *Journal of Human Evolution* **14**:203-216.

Trivers, R. L., 1971, The evolution of reciprocal altruism, *Quarterly Review of Biology* **46**:35-57.

Trivers, R. L., 1985, *Social Evolution*, Benjamin-Cummings, Menlo Park, CA.

Trouillot, M.-R., 2002, Adieu, culture: A new duty arises, in: *Anthropology beyond Culture*, R. G. Fox and B. J. King, eds., Berg, Oxford, pp. 37-60.

Tschentscher, F., C. Capelli, H. Geisert, H. Krainitzki, R. W. Schmitz et al., 2000, Mitochondrial DNA sequences from the Neanderthals, in: *Neanderthals and Modern*

Humans: Discussing the Transition, J. Orscheidt and G.-C. Weniger, eds., Neanderthal Museum, Mettmann, pp. 296-302.

Turk, I., ed. 1997a, *Mousterian "Bone Flute" and Other Finds from Divje Babe I Cave Site in Slovenia*, Znanstvenoraziskovalni Center SAZU, Ljubljana.

Turk, I., 1997b, Chronology, in: *Mousterian "Bone Flute" and Other Finds from Divje Babe I Cave Site in Slovenia*, I. Turk, ed. Znanstvenoraziskovalni Center SAZU, Ljubljana, pp. 67-72.

Turk, I., J. Dirjec and B. Kavur, 1997a, Description and explanation of the origin of the suspected bone flute, in: *Mousterian "Bone Flute" and Other Finds from Divje Babe I Cave Site in Slovenia*, I. Turk, ed. Znanstvenoraziskovalni Center SAZU, Ljubljana, pp. 157-175.

Turk, I., J. Dirjec and B. Kavur, 1997b, A-t-on trouvé en Slovénie le plus vieil instrument de musique de l'Europe? *L'Anthropologie* **101**:531-540.

Turk, I., J. Dirjec, G. Bastiani, M. Pflaum, T. Lauko et al., 2001, Nove analize "piščali" iz Divjih bab (Slovenija), *Arheološki Vestnik* **52**:25-80.

Turk, I., G. Bastiani, B. Blackwell and F. Z. Horusitzky, 2003, Putative Mousterian flute from Divje Babe I (Sloveinia): Pseudoartefact or true flute, or who made the holes, *Arheolški Vestnik* **54**:67-72.

Tylor, E. B., 1889, *Primitive Culture: Researches into the Development of Mythology, Philosophy, Religion, Language, Art, and Custom. 3rd edition*, Holt, New York.

Ungar, P. S., F. E. Grine, M. F. Teaford and A. Pérez-Pérez, 2003, A review of interproximal wear grooves on fossil hominin teeth with new evidence from Olduvai Gorge, *Archives of Oral Biology 46:285-92* **46**:285-292.

Van Schaik, C. P. and C. D. Knott, 2001, Geographic variation in tool use on *Neesia* fruits in orangutans, *American Journal of Physical Anthropology* **114**:331-342.

Van Schaik, C. P., M. Ancrenaz, G. Borgen, B. Galdikas, C. D. Knott et al., 2003, Orangutan cultures and the evolution of material culture, *Science* **299**:102-105.

Vandermeersch, B., 1970, Une sépulture moustérienne avec offrandes découverte dans la grotte de Qafzeh, *Comptes Rendues de l'Académie de Sciences, Paris* **270**:298-299.

Velarde, M. G. and C. Normand, 1980, Convection, *Scientific American* **243**:92-105.

Velo, J., 1984, Ochre as medecine: A suggestion for the archaeological record, *Current Anthropology* **25**:674.

Velo, J., 1986, The problem of ochre, *Mankind Quarterly* **26**:229-237.

Verbicky-Todd, E., 1984, *Communal Buffalo Hunting among the Plains Indians: An Ethnographic and Historic Review*, Archaeological Survey of Alberta, Occasional Paper No. 24.

Vértes, L., 1959, Churinga de Tata, *Bulletin de la Société Préhistorique Française* **56**:604-611.

Vértes, L., ed. 1964, *Tata - eine mittelpaläolitische Travertin-Seidlung in Ungarn*, Akadémia Kiadó, Budapest.

Villa, P., J. Courtin and D. Helmer, 1988, Cannibalism in Old World Prehistory. Physical Anthropology and Prehistoric Archaeology, *Rivista di Antropologia* **66**:46-64, 471-475.

Villa, P., 1992, Cannibalism in Prehistoric Europe, *Evolutionary Anthropology* **1**:93-104.

Vincent, A., 1988, L'os comme artefact au Paléolithique moyen: Principes d'étude et premiers résultats, in: *L'Homme de Neandertal, Vol. 4, La Technique.*, L. R. Binford and J.-P. Rigaud, eds., E.R.A.U.L, Liège, pp. 185-196.

Vogelsang, R., 1998, *Middle Stone Age Fundstellen in Südwest-Namibia*, Heinrich-Barth Institut, Köln.

Vygotsky, L. S., 1962, *Thought and Language*, M.I.T. Press, Cambridge, MA.

Wadley, L., B. Williamson and M. Lombard, 2004, Ochre in hafting in Middle Stone Age southern Africa: A practical role, *Antiquity* **78**.

Walker, N. J., 1987, The dating of Zimbabwean rock art, *Rock Art Research* **4**:137-149.

Wallman, J., 1992, *Aping Language*, Themes in the Social Sciences, Cambridge University Press, Cambridge.

Warner, W. L., 1937/1958, *A Black Civilization: A Social Study of an Australian Tribe (revised edition)*, Harper and Brothers, Chicago.

Watson, K. A., 1970, Neanderthal and Upper Paleolithic burial patterns: A reexamintion, *Mankind* 7:302-306.

Watts, D. P. and J. C. Mitani, 2002, Hunting behavior of chimpanzees at Ngogo, Kibale National Park, Uganda, *International Journal of Primatology* 23:1-28.

Watts, I., 2002, Ochre in the Middle Stone Age of southern Africa: Ritualized display or hide preservative? *South African Archaeological Bulletin* 57:1-14.

Weaver, A. H., T. W. Holliday, C. B. Ruff and E. Trinkaus, 2001, The fossil evidence for the evolution of human intelligences in Pleistocene *Homo*, in: *In the Mind's Eye : Multidisciplinary Approaches to the Evolution of Human Cognition*, A. Nowell, ed. International Monographs in Prehistory, Ann Arbor, pp. 154-171.

Weber, T., 2000, The Eemian *Elephas antiquus* finds with artefacts from Lehringen and Gröbern: Are they really killing sites? in: *La chasse dans la préhistoire*, C. Bellier, P. Cattelain and M. Otte, eds., Liège, pp. 177-185.

Weiss, K. and F. Hayashida, 2002, Kulturcrisis! Cultural evolution going round in circles, *Evolutionary Anthropology* 11:136-141.

Wendt, W. E., 1974, "Art mobilier" aus der Apollo 11-Grotte in Südwest-Afrika: Die ältesten datierten Kunstwerke Afrikas, *Acta Praehistorica et Archaeologica* 5:1-42.

Wendt, W. E., 1976, "Art mobilier" from the Apollo 11 Cave, South West Africa: Africa's oldest dated works of art, *South African Archaeological Bulletin* 31:5-11.

Wetzel, R. and G. Bosinski, 1969, *Die Bocksteinschmiede im Lonetal (Markung Ramingen, Kreis Ulm)*, Müller und Gräff, Stuttgart.

White, L. A., 1949, *The Science of Culture: A Study of Man and Civilization*, Farrar, Straus, and Company, New York.

White, L. A., 1959, The concept of culture, *American Anthropologist* 61:227-251.

White, L. A., 1975, *The Concept of Cultural Systems: A Key to Understanding Tribes and Nations*, Columbia University Press, New York.

White, R. K., 1982, Rethinking the Middle-Upper Paleolithic transition, *Current Anthropology* 23:169-192.

White, R. K., 1992, Beyond art: Toward an understanding of the origins of material representations in Europe, *Annual Review of Anthropology* 21:537-564.

White, R. K., 1993, A social and technological view of Aurignacian and Castelperronian personal ornaments in France, in: *El origen del hombre moderno en el suroeste de Europa*, V. Cabrera Valdés, ed. U.N.E.D, Madrid, pp.

White, R. K., 2001, Personal ornaments from the Grotte du Renne at Arcy-sur-Cure, *Athena Review* 2:41-46.

White, R. K., 2002, Observations technologiques sur les objets de parure, in: *L'Aurignacien de la Grotte du Renne: Les fouilles d'André Leroi-Gourhan à Arcy sur Cure (Yonne)*, B. Schmider, ed. C.N.R.S., Paris, pp. 257-266.

White, T. D., 1986, Cut marks on the Bodo cranium: a case of prehistoric defleshing, *American Journal of Physical Anthropology* 69:503-509.

White, T. D., 1987, Cannibals at Klasies? *Sagittarius* 2:6-9.

White, T. D. and N. Toth, 1991, The question of ritual cannibalism at Grotta Guattari, *Current Anthropology* 32:118-124.

White, T. D., 1995, *Acheulian Man in Ethiopia's Middle Awash Valley: The Implications of Cut Marks on the Bodo Cranium*, Nederlands Museum voor Anthropologie en Praehistorie, Amsterdam.

White, T. D., B. Afshaw, D. DeGusta, H. Gilbert, G. D. Richards et al., 2003, Pleistocene *Homo sapiens* from Middle Awash, Ethiopia, *Nature* 423:742-747.

Whiten, A., J. Goodall, W. C. McGrew, T. Nishida, V. Reynolds et al., 1999, Cultures in chimpanzees, *Nature* **399**:682-685.

Whitten, P. L., 1982, *Female Reproductive Strategies among Vervet Monkeys*, PhD. Dissertation, Harvard University.

Wiessner, P., 1983, Style and social information in Kalahari San projectile points, *American Antiquity* **48**:253-276.

Wiessner, P., 1984, Reconsidering the behavioral basis for style: A case study from the Kalahari San, *Journal of Anthropological Archaeology* **3**:190-243.

Wiessner, P., 1990, Is there a unity to style? in: *The Uses of Style in Archaeology*, M. W. Conkey and C. Hastorf, eds., Cambridge University Press, Cambridge, pp. 105-112.

Wilkins, J., 1998, What's in a Meme? Reflections from the perspective of the history and philosophy of evolutionary biology, Journal of Memetics 2:[http://jom-emit.cfpm.org/].

Williams, G. C. and D. C. Williams, 1957, Natural selection of individually harmful social adaptations among sibs with special reference to social insects, *Evolution* **11**:32-39.

Wilson, D. S., 1975, A theory of group selection, *Proceedings of the National Academy of Sciences* **72**:143-146.

Wilson, D. S., 1990, Weak altruism, strong group selection, *Oikos* **59**:135-140.

Wilson, D. S., 1997, Altruism and organism: Disentangling the themes of multilevel selection theory, *American Naturalist* **150**:S122-S134.

Wilson, D. S. and K. M. Kniffen, 1999, Multilevel selection and the social transmission of behavior, *Human Nature* **10**:291-310.

Wilson, D. S., 2002, *Darwin's Cathedral: Evolution, Religion, and the Nature of Society*, University of Chicago Press, Chicago.

Wilson, E. O., 1975, *Sociobiology: The New Synthesis*, Belknap Press of the Harvard University Press, Cambridge, MA.

Wind, J., 1989, Les néandertaliens ont-ils parlés? in: *L'Homme de Néandertal, Vol. 5, La Pensée.*, O. Bar Yosef, ed. E.R.A.U.L., Liège, pp. 117-124.

Wobst, H. M., 1977, Stylistic behavior and information exchange, in: *Papers for the Director: Research Essays in Honor of James B. Griffin*, C. E. Cleland, ed. pp. 317-342.

Wonderly, D. M., 1996, *The Selfish Gene Pool: An Evolutionarily Stable System*, University Press of America, Lanham, MD.

Wood, B. A., 1991, *Koobi Fora Research Project. Volume 4. Hominid Cranial Remains*, Clarendon Press, Oxford.

Wrangham, R. W., W. C. McGrew, F. B. M. de Waal and P. G. Heltne, eds., 1994, *Chimpanzee Cultures*, Harvard University Press, Cambridge, MA.

Wreschner, E. E., 1975, Ochre in prehistoric contexts. Remarks on its implications to the understanding of human behavior, *Mi'Tekufat Ha'Even* **26**:205-207.

Wreschner, E. E., 1985, Evidence and interpretation of red ochre in the early prehistoric sequences, in: *Hominid Evolution: Past, Present and Future*, P. V. Tobias, ed. Alan R. Liss, New York, pp. 387-394.

Wrinn, P. J. and W. J. Rink, 2003, ESR dating of tooth enamel from Aterian levels at Mughâret el 'Aliya (Tangier, Morocco), *Journal of Archaeological Science* **30**:123-133.

Wuketits, F. M., 1996, Biosocial determinants in moral behavior: An evolutionary approach, *Homo* **46**:113-124.

Wurz, S., 1999, The Howiesons Poort backed artefacts from Klasies River: An argument for symbolic behavior, *South African Archaeological Bulletin* **54**:38-50.

Wynn, T. G., 1989, *The Evolution of Spatial Competence*, Illinois Studies in Anthropology 17, University of Illinois Press, Urbana.

Wynn, T. G. and F. Tierson, 1990, Regional comparison of shapes of later Acheulean handaxes, *American Anthropologist* **92**:73-84.

Wynn, T. G., 1991, Tools, grammar and the archaeology of cognition, *Cambridge Archaeological Journal* **1**:191-206.

Wynn, T. G., 1993, Layers of thinking in tool behavior, in: *Tools, Language and Cognition in Human Evolution*, K. R. Gibson and T. Ingold, eds., Cambridge University Press, Cambridge, pp. 389-406.

Wynn, T. G., 1996, The evolution of tools and symbolic behavior, in: *Handbook of Human Symbolic Evolution*, A. Lock and C. R. Peters, eds., Clarendon Press, Oxford, pp. 263-287.

Wynn, T. G. and F. L. Coolidge, 2004, The expert Neandertal mind, *Journal of Human Evolution* **46**:467-487.

Wynne-Edwards, V. C., 1962, *Animal Dispersion in Relation to Social Behavior*, Oliver and Boyd, Edinburgh.

Wynne-Edwards, V. C., 1986, *Evolution through Group Selection*, Blackwell Scientific Publications, Oxford.

Youngs, B. S., 1810 *The Testimony of Christ's Second Appearing: Containing a General Statement of All Things Pertaining to the Faith and Practice of the Church of God in This Latter Day. Second Edition*, E. and E. Hosford, Albany, NY.

Zilhão, J., 2000, The Ebro frontier: A model for the late extinction of Iberian Neanderthals, in: *Neanderthals on the Edge. Papers from a Conference Marking the 150th Anniversary of the Forbes' Quarry Discovery, Gibraltar*, C. B. Stringer, R. N. E. Barton and J. C. Finlayson, eds., Oxbow Books, Oxford, pp. 111-121.

Zilhão, J. and F. D'Errico, 2000, La nouvelle « bataille aurignacienne » Une révision critique de la chronologie du Châtelperronien et de l'Aurignacien, *L'Anthropologie* **104**:17-50.

Zilhão, J., 2001, *Anatomically Archaic, Behaviorally Modern: The Last Neanderthals and Their Destiny*, Nederlands Museum voor Anthropologie en Praehistorie, Amsterdam.

INDEX